Studies in Fuzziness and Soft Computing

Volume 348

Series editor

Janusz Kacprzyk, Polish Academy of Sciences, Warsaw, Poland
e-mail: kacprzyk@ibspan.waw.pl

About this Series

The series "Studies in Fuzziness and Soft Computing" contains publications on various topics in the area of soft computing, which include fuzzy sets, rough sets, neural networks, evolutionary computation, probabilistic and evidential reasoning, multi-valued logic, and related fields. The publications within "Studies in Fuzziness and Soft Computing" are primarily monographs and edited volumes. They cover significant recent developments in the field, both of a foundational and applicable character. An important feature of the series is its short publication time and world-wide distribution. This permits a rapid and broad dissemination of research results.

More information about this series at http://www.springer.com/series/2941

Jordi Cat

Fuzzy Pictures
as Philosophical Problem
and Scientific Practice

A Study of Visual Vagueness

 Springer

Jordi Cat
Department of History
    and Philosophy of Science
Indiana University
Bloomington
USA

ISSN 1434-9922   ISSN 1860-0808   (electronic)
Studies in Fuzziness and Soft Computing
ISBN 978-3-319-83674-4   ISBN 978-3-319-47190-7   (eBook)
DOI 10.1007/978-3-319-47190-7

Printed on acid-free paper

This Springer imprint is published by Springer Nature
The registered company is Springer International Publishing AG
The registered company address is: Gewerbestrasse 11, 6330 Cham, Switzerland

*In memory of Patrick Suppes
and Richard Wollheim*

Acknowledgments

I am grateful to Cathy Helgason for having introduced me to her research in the medical application of fuzzy set theory, especially after I had studied the methodological role of imprecision and uncertainty in Otto Neurath's philosophy of science. I am also indebted to Enric Trillas and Rudolf Seising for their continued patience, tolerance, generosity and encouragement. An earlier version of the content of Part II was published in the August and December 2015 numbers of *Archives for the Philosophy and History of Soft Computing*; I wish to thank the editors for permission to reproduce the material that I haven't revised or developed further in this book. Finally, I am indebted to George Clerk for help with the photographs and Omar Sosa Tzec for help with the diagrams.

Contents

List of Figures

Part 1

Part I

Chapter 1
Introduction: Visual Uncertainty, Categorization, Objectivity and Practices and Values of Imprecision

Visual perception and depiction involve the experience of fuzzy images. In this book I aim to understand this rich phenomenon, so pervasive in different cognitive situations and with a long history of pictorial practices, some more practical, others more creative. My strategy is to adopt a broad perspective to examine the phenomenon in relation to philosophical problems and scientific and technological practices; in this matter, imaging philosophy and imaging sciences provide each other with challenges and resources. Philosophical accounts of language and scientific models of image processing leave us with a shared conceptual space in which to raise the questions, how is vagueness in images possible? How is it unavoidable or at least effective? What values, meanings or uses does it have? In this book I seek to provide answers to these questions that touch on five connected issues: empiricism, vagueness, representation, application of mathematics and the relation between philosophy and science. In the process I provide a non-technical conceptual discussion of the issues, concluding with a few challenges and suggestions for the mathematical modeling, application and manipulation of fuzzy images.

Reports of fuzziness are commonplace as well as multifaceted. But in varying degrees they share three kinds of claims: We see fuzzily, we see fuzzy things and we see fuzzy pictures. We also see sharply, see sharp things and see sharp pictures. The approach I lay out below seeks to accommodate these three types of experiences. The vocabulary of the phenomenology and its representation is rich and pervasive: haziness, blurring, slurring, fuzziness, indefiniteness, unclearness, ghosting, distortion, confusion, diffusion, dispersion, irresolution, indistinctness, approximation, uncertainty, etc. In addition, we process digital images to treat their fuzziness and enable their identification. Any consideration of pictorial and visualization technologies and traditions, from classical painting and photography to assisted vision and digital imaging, suggests that sharpness and fuzziness, precision and imprecision are understood and treated differently in different contexts, always relative to varying standards, values and background knowledge.

© Springer International Publishing AG 2017
J. Cat, *Fuzzy Pictures as Philosophical Problem and Scientific Practice*,
Studies in Fuzziness and Soft Computing 348,
DOI 10.1007/978-3-319-47190-7_1

The value of experience is in itself also significant. Empiricism is the philosophical commitment to a fundamental role of experience in pursuing the goals of inquiry, including the goals of scientific methodology in empirical domains. But in order to sound reasonable and credible this view must be qualified in a number of ways. The value of experience cannot be reduced to the role of perception. We must take into consideration that experience takes the form of different processes of causal interaction with the world, different kinds of products of those interactions, different processes to treat those products, and different roles and values those products might get in the pursuit of different scientific aims. The use of all sorts of technological and methodological controls introduces complexities that enable but also limit the role of experience, its reliability, its interpretation, its evidentiary role, etc. I am referring to the values of experiments, data and visualizations.

Even in science, then, especially in science, perception cannot be reduced to seeing or looking as processes and activities involving only one sense modality. But it has become a dominating modality in science and in modern life. Technologically, instruments for seeing or looking, ever more powerful and discriminating, have helped establish this dominance of the visual, even when they primarily involve visualizing modes of detection that connect with our eyes only distantly. The ongoing project of computer vision, fuzzy and sharp, is an extreme version of the use of technology to go beyond computational modeling of perception and to simulate and replace organic visual mechanisms.

My discussion considers images of two kinds that are related but different. On the one hand, I refer to images as perceptions, specifically as visual mental representations or the controversial phenomenological or mental content of perception. On the other, I refer to images as external, publicly accessible, objective material constructions or uses, the kind of representation or means of communication we call pictures. Each kind presents peculiarities that inform the different uses and meanings of their imprecision and also of its mathematical representation. Their relation may be complex but is evident in the way pictures may themselves be considered content of visual experience and in the role visual experience plays in the process of their design, production or use—including, in the case of representations, the visual experience of their possible intended or adopted content. My discussion, then, focuses on what I call *pictorial empiricism*: the methodological attitude that (1) makes sense of the role of pictures as an objective extension of experience and extends to their role in scientific practice; and that (2) makes room for, as well as makes sense of, a variety of cases of fuzziness and their varied significance. I want to identify the epistemic and methodological place of fuzziness in scientific pictures (or in the scientific treatment of images).

My concern with the scientific dimension of fuzziness is inseparable from a broader attention to the values of imprecision, vagueness, or uncertainty. The main premise of my analysis is the recognition that fuzziness is widely unavoidable and unreasonably effective. For instance, we find visual representations of visible systems realistic or naturalistic on account of their incompleteness and fuzziness; those restrictions on visual information mimic our visual experience. The use of perspective is another. The perceptual standard of accuracy has sometimes been called the subjective standard of truth. What the standard amounts to has its own history and controversies.[1] The same partial, selective aspect that characterizes our perceptual attention also characterizes scientific modeling; only in that case the partial, selective aspect characterizes a rather more conceptual or theoretical focus and perspective.

In different contexts, the particular value or significance of fuzziness will vary. From my broader perspective, it is the result of its place in practices and its relation to the considerations that constrain them. Some cognitive practices have more generic functionality, others develop in more restricted contexts such as the visual arts. In other contexts fuzziness in the determination of perceptual and less perceptual content plays a role in establishing visual findings and evidence; it plays a role in visual analysis in areas as diverse as chemical spectroscopy, engineering, microbiology, biomedicine, geology, astrophysics, archeology, ecology, forensic science, civil and military intelligence and ethical and legal analysis.

In the case of the sciences, the diverse significance of fuzziness results from its place in different applications and interpretations of formal and technological practices, from quantitative measurement to image processing. We might note more generally that the place of fuzziness is twofold: as a tool in scientific practices aimed at inquiry and intervention—for instance, in models of expert identification, data and evidence—and as the subject of scientific representation and intervention. The second role of fuzziness makes contact with other practices of visual perception and depiction. Here, of course, I shall emphasize that the second role is now mainly associated with the application of the formalism of fuzzy sets, and that should be used in effect to contribute to the first role.

The application of mathematics is itself a formal sort of technological control that constraints, determines, clarifies and helps apply the idea of vagueness in more limited but also more powerful ways. This is a distinctive value of mathematics and its application. Its rules and standards help control the meaning and computation of results and the validity of proofs. In other words, such controlled and controlling features reveal mathematics as a complex normative practice. Fuzzy set theory provides a similar sort of benefit through a formal determination of vagueness that helps represent and apply vague representations and reasoning as objective and regulated instruments.

How is vagueness conceptually possible and empirically actual, pervasive and unreasonably effective, embedded in value judgments and practices? The broader perspective I adopt in the book locates vagueness in judgments of categorization and its indeterminacy. I rely on an emphasis on categorization for different reasons. One is practical, to explore and develop one single approach; another is methodological, its adequacy and generality in understanding cognitive practices and, most importantly, its specific role in in two formal standards: philosophical models of linguistic vagueness and the application of mathematical fuzzy set models. By focusing on pictures and depiction I seek to explore and extend this collaboration to

---

[1]Gombrich [6], Gibson [5].

certain conclusions. My discussion relies on the fact that philosophical and mathematical concepts have overlapped in the domain of semantic and logical practices involving predication and its truth. But pictures aren't words. Thus, in order to extend this framework, I adopt a set of extended notions that examine and problematize the difference and relation between understanding the modeling of fuzziness in words and pictures.

The broader perspective helps identify key similarities, differences and joint functions in interaction. Again, the approach aims to accommodate representation in perception and depiction, internal images and external pictures, as they combine in ordinary and scientific practices of representing and intervening. One consequence is that the approach requires a broader notion of objectivity that bridges the gap between the epistemic and the objective philosophical interpretations of vagueness and between the philosophical accounts and mathematical practices. There I can locate vagueness as a practice and in practice: in an objective world of both things and cognitive practices that include categories and categorization, predicates and properties, pictures and depiction, representations and representing.

The generality of this kind of objective account bypasses the terms of a philosophical debate between opposed interpretations of vagueness, epistemic and objective. According to epistemic accounts, vagueness is a cognitive feature of language users, typically a cognitive deficiency. According to objective accounts, vagueness is a semantic feature of linguistic predicates that refers to the corresponding properties of a thing as described and that make (partially) true statements about them in terms of those predicates. The purely semantic standard of truth in terms of individual properties corresponding to predicates used in the categorization becomes a further conceptual standard of objectivity that might still apply, but only alongside further metaphysical commitments. By contrast, the minimally objective content of the predicate, precise or imprecise, when applied to a particular thing, event or situation derives its objectivity from three relations: this thing's relation to others, to the specific practices of categorization and to the context of their exercise, that is, the conditions of application of the relevant standards and constraints. This is the place where much philosophical literature has identified the phenomena of framework dependence and fact-value entanglement. Elsewhere I have already discussed the subjective elements of applying fuzzy predicates and how applying the formalism of set theory provides their objective dimension.[2] The epistemic, ontic and normative (also pragmatic) dimensions of categorization are deeply entangled. Along the same lines, my broader perspective can also accommodate the ambiguity in the contemporary use of related concepts such as the epistemically objective notions of information and probability. Naturally, one serious challenge consists in sorting out the methodological relation between the concepts of uncertainty, noise and error.

To understand vagueness in this complex domain of situations and activities, I introduce several distinctions: between perception and depiction; between three

---

[2]Cat [2].

kinds of contents, *intrinsic, extrinsic* and *external* (target), where the intrinsic is the most basically associated with perceptual properties; also between two kinds of indeterminacy and various relations between them. While such distinctions are themselves contextual, my framework goes beyond the distinction, equally hard to draw, between sensory and meaningful perception in the literature on perception.[3]

Of course, since I am placing categorization, representation and the uses of fuzziness at the center of the argument that connects linguistic and pictorial vagueness and philosophical and scientific approaches, I rely for illustration and evidence on examples from different kinds of practices. On matters of representation and its links to perception and language, my sources are quite multi-disciplinary. For instance, I point to cognitive science, set theory, imaging sciences, aesthetics and the history of painting and photography (especially from works and reflections from Italian Renaissance painting, nineteenth-century Victorian painting and photography and contemporary German painting). Such practices extend the kinds of constraints that apply to perception, and they do so, I believe, independently of the details postulated by different explanatory mechanisms. With such constraints in mind, I will point to particular constraints that inform technological practices of image processing. The role of historical evidence is not just to document significant phenomena of fuzziness; it's also to document interpretive problems and prescriptive standards. Interestingly, we can find different attitudes to the significance of fuzziness, each of which is embedded in practices it also guides. The most familiar one is the pervasive high-definition standard that in different contexts and for different purposes—from entertainment to security and medical diagnostics—seeks to eliminate fuzziness through technical processing.

In ordinary and technical discussion of images, it is common to identify mental images, perceptions and pictures, with internal or external representations. The internal/external distinction is also assumed in the distinction between epistemic and objective accounts of vagueness. In the linguistic case, predicates and propositions are the internal representations standing in a semantic relation to some outside world with objective reality. When we focus on knowledge from the same perspective, going back to Plato, we are told that a subjective mental world contains beliefs that play the role of more or less accurate internal representations of reality. The relations between those representations and empirical data or else fundamental principles characterize the justification and revision of beliefs in realist and rationalist theories of knowledge.

An alternative perspective is the so-called constructivist-interactionist accounts.[4] They emphasize the constructive role of cognition in interaction between embodied, situated cognitive agents and their environment. On this kind of account, beliefs and other particular conditions of inquiry constrain the determination of grounds for forming and accepting new beliefs. What beliefs are, how they are related to the

---

[3]Gong et al. [7].

[4]For the distinction and defenses of this kind of account of ordinary and scientific knowledge see Smith [11].

world and how they change are explained by how they are part of the world, that is, in terms of actions and interactions; and these are often described with evolutionary and ecological analogies.[5]

This approach to cognition relies on dynamical models of reciprocal action and determination. But thinking and acting in terms of any interaction assumes the identification of the distinct systems that engage in such a relation—individuation and existence claims aside. And the distinction implies a conceptual boundary that belongs within a range of dualities such as internal/external, mind/body, will/action, word/world, subject/object, organism/environment, nature/culture and cause/effect. But where does the environment start? Where does the body end? Etc. Like the notion of interaction, these questions are hardly intelligible within a holistic continuous picture without boundaries. Alternatively, we can think of the distinctions between agent and environment, word and object, etc. as themselves constructions, more or less stable precisifications of boundaries that are not arbitrary; rather, they are part of the dynamics of interaction and reciprocal modification and is related to a context of the agent's actions informed by goals and constraints.[6]

The role of beliefs, even as a cognitive hypothesis, like the use of linguistic statements raises the specter of truth-talk. From a functional standpoint, truth supports understanding, reasoning and acting. The interpretation of statements and the validity of inferences seem controllable and intelligible only by appealing to an ideal of truth. Rational planning and decision-making seems to depend on the kind of information we can rely on. But is it truth? What sort of truth?

Categorization is reflected in statements where it is encoded in predication, but truth only applies to the use of those statements. To the business at hand, truth has long been considered irrelevant to vagueness in predication, and its cognitive role, if any, fails to bridge any semantic gap that realist or objectivist dualisms have long associated with precision, for instance beginning with the formal objectivity endorsing the trustworthiness of mathematical concepts and reasoning. No numbers, no facts; no precision, no truth.[7] Anything else is only a matter of inaccuracy or degree of error and approximation. In different forms, the issue runs through the entire book. By the end, five questions should look relevant and remain

---

[5]Cognitive agents are said, for instance, to occupy a niche.

[6]In this contextual way determined by functions, goals and structural and other constraints, the naturalized hermeneutic circle becomes virtuous. The precisification also illustrates part of the ideas about vagueness of categorization in this book. The move helps make sense of statements such as the following: 'while our individual perceptual and behavioral tendencies are certainly shaped by, among other things, the history of the species' more or less effective interactions with its environment, the reciprocal of this also obtains. That is, the features of the environment, sabre-toothed tigers included, with which we and past members of the species could (have) interact (ed) have never been independent of out particular structures and how we were already operating as perceiving and behaving organisms. In other words, the ways in which we have evolved have depended at every point on what we already were, which means also what we believed, as well as vice versa. Reciprocal determination of this kind is a central mechanism in the dynamics of cognition (…).' Ibid., 49.

[7]Pereboom [9], Porter [10], Wise [12].

unanswered: Why truth? What truth? Does vagueness require truth? Is there a scientific standard of truth? Do the interpretation, justification and application of fuzzy set theory require any notion of truth?

My guiding assumption is that truth-telling, like categorization, reference to facts and discovery or the application of mathematics, is a kind of practice embedded in contexts of inquiry and communication that are constrained by multiple norms and activities—formal, theoretical, material, methodological, institutional, social, etc.[8] If truth-talk is a way of categorizing, representing or communicating information or prompting inquiry, it is truth enough. Many aims and activities in our lives and in the sciences, including fuzzy set theory, might support but do not demand the application of a truth as a fixed semantic relation between models or their symbolic presentation and some independent, prior "real" world. They make do with aiming at, for instance, cognitive utility—in the form of empirical support, reasonability, informativeness, accuracy, consistency, unity, simplicity, tractability, robustness, predictive power, etc.—and practical relevance. Both the adoption of the respective criteria and the conditions of their application require a rich cognitive and social context. Still, explanatory causes and other properties, entities and phenomena may extend our conceptual and practical resources—beyond the domain of data and facts we trust and seek to understand. But to endorse them as an aim, along with the semantic criterion and the ontic dualism that supports it, is to express a particular understanding and motivation of science. Truth-talk is the eye in the tornado of scientific practices.

Internal representations called beliefs can be understood as configurations of perceptual and behavioral tendencies of organisms entangled in an environment that they reshape and "represent." New beliefs are formed or at least stabilized in pragmatic terms, as having more reliable and functional consequences of the actions they initiate and guide relative to interests and other constraints.[9] The relation to the world and the dynamics of their change are the same: 'the continuous mutually shaping interactions between our more or less changing beliefs and the more or les changing worlds in which we operate with them.'[10] From this dynamical, inter-active, reciprocal standpoint, truth—veridicality, etc.—is interchangeable with conceptual stability, pragmatic reliability and serviceability—achieving evidential support, consensus, etc. To many this view will strike as an expression of anti-representationalism. But we are free to pretend otherwise and impose on this picture any standard of representation, dualism and semantic constraints that we find possible, serviceable and intelligible in light of out capacities and limitations. Not any truth will do.[11] And this includes a minimal model-theoretic standards of truth: that a set of objects—by our own lights—is a truth-maker for a statement or symbolic relation provided we can map its members onto the extension and

---

[8]Elgin [4], Cat [3].
[9]Smith [11, Chap. 3].
[10]Ibid., 43.
[11]Elgin [4, Chap. 1].

structure of the terms in the statement or equation; the set is a model of the symbolic relation that it satisfies.

I adopt this perspective, instead, to assume minimal notions that can assimilate, rather than dismiss, also the traditional vocabulary of visual representation without reducing it to ontic dualisms. It is a heuristic to situate the role of categorization in relation to words and pictures and their scientific modeling. Categorizations are dynamically stable cognitive tendencies that organize interactions and behaviors in relation to the environment according to dynamical patterns with contextual fit. Representation is the activity and the product of fitting constructs according to an instrumental notion of truthfulness, match or fit as practical and functional forms of fitness.[12] Whatever dualities and labels we introduce are only formal, abstract ways of tracking relatively stable and effective uses. We may call pictures and perceptions representations then insofar as they represent in this minimal interactive, contextual sense. In other words, a representation is an activity and a function—and not an object—that involve perceptual traces of the relevant acts; these traces in turn are presented as the object-representation, the visual mark—word or picture—in a material support or medium of production and display.

Categorization and representation, content and meaning, may be effects. They may be better construed as ways of interpreting and participating that like linguistic rules and other conventions occur in what in linguistic communication has been called a system of "reciprocal effectivity": dimensions of patterns of "interlooping acts and reactions" that succeed in affecting the behavior of participating speakers through changes in their beliefs, feelings, habits and reflex associations.[13] Game-theory models of animal communication through signaling illustrate this perspective.

My account in terms of the fit and objectivity of the content of representations is minimally of the same kind, allowing for—without making—an additional commitment to the classical realist ontology that supports the semantic distinction between language and prior and autonomous objects that words might denote and provide the objective conditions that make statements true or false.[14] For the case of vague language this is the assumption of ontic objectivity for fuzzy properties and about the partial "truth" of propositions and pictures with vague predicates and fuzzy images representing those properties (see below). It is an empirical matter how the use of language and semantic distinctions above have emerged and been used stably within linguistic communities of interacting agents. It is also a more theoretical matter whether the semantic accounts developed in those terms are best

---

[12]Some beliefs and categorizations are about belief and representation themselves, and my view, while limited in scope and application, should be independent of them.

[13]See Smith [11, Chap. 4] for a discussion of communication without determinate content or meaning making sense of the introduction of habits of recording and enforcing norms and meanings as mere markers for pre-existing stable patterns of effective coordination between verbal agents. This is a naturalistic version of views in philosophical doctrines such as Merleau-Ponty's phenomenology; see Merleau-Ponty [8].

[14]Ibid.

revised or rejected. In general, different specific conceptual assumptions, formal tools and empirical situations suggest further developments and alternatives when understanding and applying fuzzy set theory.

My focus on categorization and representation is not a commitment to their relation being either fundamental or exclusive—or to an underlying ontic view. I will assume that categorization is not only relevant to representing, just as representing itself might not depend on categorization alone, or even at all. Neither is representation the main use of images; it may play a role in other goal-oriented, problem-solving practical tasks or enable other salient effects such as trigger a mechanical or emotional reaction; but it may as well be irrelevant to achieving them. Of course, as an explanatory matter, we might always be in a position to postulate undetected activities of categorization for the sake of explanation, just like we can postulate the operation of opaque rules.[15]

The same opaqueness extends to the relation between categorization and recognition. Recognition is typically understood as a phenomenon or activity of identification, which in turn involves either or both of two related tasks: (1) judging a particular object, situation, event, a pattern, etc., as identical with another, as numerically the same; and (2) judging something as being of a particular kind, which my be considered as having a property or being a member of a group, class, etc. The second kind of judgment is what we call categorization or classification; it may depend on establishing a similarity to others things in some respect. The similarity judgment is not one of numerical identity; it is comparative. In this regard, the second kind is related to the first, as part of the assessment of identity. Still, the first kind may be partly explained also in terms of the second, by positing a transparent or an opaque mechanism involving categorization. The entanglement cannot be reduced to mere definition of categorization in terms of recognition or vice versa.

I conclude that a shared framework centered on categorization offers suggestions on four related philosophical topics: pictorial empiricism, vagueness, representation and mathematical application. Pictorial vagueness differs from linguistic vagueness in interesting ways. Two senses of 'representation' are related: picture and depiction. Being a vague picture is different from providing vague depiction. Vague pictorial representation is not reducible to truth of vague predication by the picture. Also, vague pictures, unlike vague predicates, may be themselves independently vague. So may be their subject matter and also their target. The relation between vague representations and vague representing is complex. The picture, its contents and the target's independent background, and perceptual (or pictorial) representation may present different states of determinacy and relate to one another in contextual ways that are neither unique nor general.

The application of mathematics to the previous results about pictorial vagueness yields interesting lessons. Specifically, fuzzy modeling offers a unified framework

---

[15]The scope of opaque cognition we find acceptable becomes a matter of methodological and epistemic commitments such as the role of simplicity and standards of explanation.

and precedent for treating vagueness in words and images. But unity of application isn't simple. I have mentioned above that I want to problematize the relation between the different applications. I will claim, for instance, that fuzzy pictorial representation is not reducible to simple relations of partial truth and fuzzy predication. Categorization, I suggest, is not an exclusively linguistic activity. It may equally support pictorial and symbolic representations, although it is typically with the latter that we associate the cognitive tasks of representation, determination and efficient communication.

The relation between two kinds of pictorial contents in fuzzy representation hints at the complexity of mathematical application and how it can be understood and motivated, as a richer practice involving formal and technological processing. So, the development and application of the fuzzy-set formalism are part of a complex practice of representing and manipulating vagueness in practice.

Connectedly, I will argue that unlike accuracy of measurement or depiction, in fuzzy-set models vagueness becomes a form of approximation, without collapsing into a form of inaccuracy, and subject to the multifaceted contextual character of more familiar approximation practices. These formal models are models of uncertainty or approximation in objective and epistemic conditions. But fuzziness understood generally in those terms should be distinguished from conceptual skepticism, we cannot tell what something is, its correct categorization.

Attention to approximation also contributes to my focus on the application of fuzzy set theory as a contextual practice, bringing out some of its aspects as an expression of the practice of approximating. Fuzziness in pictures, then, adds another dimension of approximation; and vice versa, approximation adds another dimension to the analysis of fuzziness.

Another kind of model of uncertainty and approximation is probabilistic. But classical probability theory is based on a notion of randomness applied to events that is restricted by the structure of Boolean algebras. The conceptual issue becomes whether probabilistic uncertainty can model fuzziness or vagueness. If, arguably, it cannot, the next issue is whether probabilistic or statistical models may be generalized adequately by generalizing their set-theoretic foundations that model the kinds of categorizations at work in their empirical application. The generalization might be argued to undermine the probabilistic character of the models, and this might not be as harmless an approximation as the way it also undermines their classical set-theoretic and conceptual precision. A more general notion of uncertainty is characterized, for instance, by so-called fuzzy measures that relax the additivity property of probabilities for disjoint events. This is the basis for Zadeh's possibility theory. I will point to these contrasts and possible approximations in the realm of uncertainty to suggest the relevance of modifications and replacements of statistical models. Notice that also these kinds of models admit, like vagueness, of objective and epistemic interpretations.

Categorization plays a role also in complex kinds of cognitive activities such as computation and reasoning. In typical linguistic or symbolic cases, predication and rules of inference, logical or mathematical, provide tools for reasoning in order to extend the scope of relevant information or the set of beliefs and to assess the strength of their reliability. With suitable models for the application of fuzzy predicates—for instance, models of membership functions—and logical connectives—or associated algebraic operations—, we can obtained models of approximate symbolic reasoning with rules of inference and measures of their reliability in conditions of uncertainty, or vagueness. As before, approximate reasoning faces the challenge of providing the basis for interesting approximations to statistical or probabilistic modes of inference such as Bayesian methods. Reasoning may extend representation, just like representation may aid reasoning. What about thinking? Thinking more generally involves the exercise of cognitive skills in constrained contexts; the relevant tasks aim, for instance, to provide interpretation (by some criterion of understanding), to extract information or to solve problems.

Visual thinking may then be extended to thinking in fuzzy cases in conditions of visual uncertainty. Notice the tricky character of the distinction between symbols and pictures as symbols are themselves visible marks. To the extent that also pictorial representation and thinking or reasoning are inseparable—for instance, computational models of vision–, fuzzy representation may be said to involve fuzzy thinking or reasoning. Visual thinking may involve visual perception of visible situations in general or, specifically, visual depictions. Thinking with different sorts of visual representations, subject to categorization, involves, as in the symbolic case, the capacity to carry out cognitive tasks of inference, interpretation, problem-solving, etc. Geometrical reasoning and analogical thinking in pictorial terms are familiar examples, they may serve the purposes of categorization, or the cognitive and practical uses of maps, graphs and other diagrams.

Fuzziness enters the role of categorization through thresholds of uncertainty, indistinguishability or indiscrimination. But categorization is not, in principle, a linguistic or computational affair. As a consequence, several kinds of challenges follow: the empirical challenge of modeling such processes, the technological challenge of rendering such processes automatic and the more philosophical, but related challenge, of understanding the application of fuzzy set formalism in the light of differences between visual and linguistic representations.

In this book I focus on the case of images, but the philosophical notions and issues apply equally, with the corresponding qualifications, to the use and phenomenology of other sense modalities such as hearing. A different but related analysis may be offered about sound recognition, especially speech recognition. Even in the cultural and artistic context, acoustic imprecision had aesthetic meaning before twentieth-century music. This is hardly surprising from the standpoint of both representationalist and anti-representationalist approaches to music. When art was driven by aesthetic projects such as German Romanticism, it kept a close philosophical relation to German Idealism and the epistemic problem of the role of experience in knowledge of objective reality and especially transcendent reality beyond the bounds of the finite, quantifiable intuitive domain, an immeasurable kind of spiritual and ideal reality associated with the infinite. The imprecision that Enlightenment rationalists had decried in instrumental music as a language for depicting natural events, ideas and emotions, at least their form, became soon a virtue, a vehicle of truth, in relation to the transcendent new domain.[16]

Rather than a technical survey or analysis, my aim in this book is to provide a selective philosophical and empirical discussion of philosophical problems and their scientific solutions, while identifying a plurality of contexts, meanings, uses and practices around the language of imprecision and its application to pictures.

As part of this plurality I draw attention to several issues:

(1) I point to disciplinary and cross-disciplinary plurality. The scientific solutions themselves raise new methodological and conceptual issues that deserve further investigation. In fact, the formal mathematical practices of defining the concept identified by the family of idioms that includes uncertainty, imprecision and vagueness, are carried out by engaging resources of other empirical areas of research and areas in philosophy, especially in a fuzzy overlap between philosophy and the sciences.

(2) There exist a variety of technical meanings and uses introduced in the scientific context.

(3) The task of distinguishing between linguistic and pictorial images and meanings is surrounded by values and challenges that must accommodate the relevance of their complex interaction.

(4) The language-picture, description-depiction, symbolic-visual dichotomies raise the issue of the possibility of distinct semantic relations of correspondence, accuracy or fit. This is the problem of the limits of strict criteria of linguistic truth and the challenge of a unified semantic interpretation of fuzzy set theory.

(5) I introduce a contextual distinction between fuzziness and sharpness and between precision and imprecision, and claim that in the visual context, fuzziness is always relative and contextual, as is the standard of sharpness.

(6) Within the plurality of images, I try to distinguish and relate images as perceptions (and so-called mental images) and pictures as special kinds of perceived objects. This leads to the relation between different kinds of fuzzy experiences and uses.

(7) Pictures have many uses.

(8) Finally, I emphasize the contextual distinction between two kinds of categorizations, namely, intrinsic and extrinsic content and maps, also the restricted use of specific rules connecting them supported by empirical and conceptual assumptions, purposes and standards that identify the practice and its context. The distinction is relevant to understanding the production and plural uses of pictures.

---

[16]As part of a rich and nuanced discussion of the changing aesthetics and politics that ushered in a new role for music, Bonds provides an insightful discussion of the correspondingly changing meaning of musical vagueness at the turn of the nineteenth century; see [1, Chap. 1].

The connections between these different elements of plurality structure the rest of my discussion, below, of the practices around visual fuzziness. From this broad perspective, we can identify challenges and opportunities for understanding and carrying out scientific practices of fuzzy imaging.

The structure of the book has the form of a patchwork of overlapping discussions of several connected themes indicated above; it is not a linear historical narrative, a collection of episodes, an axiomatic system, a long proof, or a pedagogical sequence of illustrated notions and results in increasing order of connected complexity. But discussions in later chapters rely on discussions in earlier ones. Readers might skip parts or read others in isolation only at the risk of their own frustration or disappointment. I also expect some readers with technical expertise to jump quickly to consider some of my observations as either mistakes or platitudes, sometimes with good reason. Some of the claims demand detail, illustration and technical development. I can only hope that my critical, broader perspective on technical conceptions and applications of different representations of imprecision will contribute to their understanding and application.

References

1. Bonds, M. E. (2006). *Music as thought*. Princeton, NJ: Princeton University Press.
2. Cat, J. (2015). An informal meditation on empiricism and approximation in fuzzy logic and fuzzy set theory: Between subjectivity and normativity. In R. Seising, E. Trillas, & J. Kacprzyk (Eds.), *Fuzzy logic: Towards the future* (pp. 179–234). Berlin: Springer.
3. Cat, J. (2016). The performative construction of natural kinds: Mathematical application as practice. In C. Kendig (Ed.), *Natural kinds and classification in scientific practice* (pp. 87–105). Abingdon and New York: Routledge.
4. Elgin, C. Z. (1997). *Between the absolute and the arbitrary*. Ithaca, NY: Cornell University Press.
5. Gibson, J. J. (1979). *The ecological approach to visual perception*. Boston: Houghton Mifflin.
6. Gombrich, E. H. (1980). Standards of truth: The arrested image and the moving eye. In W. J. T. Mitchell (Ed.), *The language of images* (pp. 181–218). Chicago: The University of Chicago Press.
7. Gong, S., McKenna, S. J., & Psarrou, A. (2000). *Dynamic vision*. London: Imperial College Press.
8. Merleau-Ponty, M. (1965). *Phenomenology of perception*. London: Routledge & Kegan Paul.
9. Pereboom, D. (1991). Mathematical expressibility, perceptual relativity, and secondary qualities. *Studies in the History and Philosophy of Science, 22*(1), 63–88.
10. Porter, T. (1995). *Trust in numbers*. Princeton, NJ: Princeton University Press.
11. Smith, B. H. (1997). *Belief and resistance*. Cambridge, MA: Harvard University Press.
12. Wise, M. N. (Ed.). (1995). *The values of precision*. Princeton, NJ: Princeton University Press.

Chapter 2
From Ordinary to Mathematical Categorization in the Visual World... of Words, Pictures and Practices

2.1 Categorization and Linguistic Representation

My initial strategy is to focus on categorization and its indeterminacy, since they play a role in understanding and using pictures as well as they do in symbolic representation and reasoning. The methodological choice to emphasize categorization is not arbitrary; it is rather motivated by the central role it plays in conceptual and formal accounts of the phenomenon of vagueness as well as in the very practice of its investigation, particularly in philosophical models of linguistic vagueness and the application of formal models of fuzzy sets. I take it to be a stable description of a phenomenon and a practice itself used to guide and explain other forms of behavior. There is some gain in trying to apply the conceptual standards developed around the linguistic cases: it brings out the scope of their relevance and some of their differences in role and interpretation. But, again, this is not to deny the variety of additional or alternative criteria relevant for understanding or establishing representation, or the mechanisms for effecting it; similarly, in the specific case of categorization, for its possible criteria or procedures. The caveat, I repeat, is that, like representation, the role of categorization itself as a label for a cognitive practice and an explanation for cognitive and other practices may very well be reinterpreted, explained away and replaced; that is, in objectivist terms, talk of categorization could be "wrong" and in pragmatic, functional terms, it could prove less efficient than alternatives. Vagueness and fuzzy set theory will have to be reinterpreted accordingly. Either way, applying specific views to new models of specific cognitive phenomena will suggest further developments that I leave the reader to explore.

Symbolic representation in a language refers and describes; that is, we have come to form beliefs about performing such functions and how to understand the behavior in such terms for the sake of communication. On the view I adopt here, linguistic tasks are performed through labeling and categorization. Eligible

J. Cat, *Fuzzy Pictures as Philosophical Problem and Scientific Practice*, Studies in Fuzziness and Soft Computing 348, DOI 10.1007/978-3-319-47190-7_2

objects of reference and description range widely from individual worldly facts—including bodily states and private intentional objects–, events, situations and things, to properties, groupings, fictional entities, language itself, etc. Indeed, categorization attributes kinds or qualities for descriptive and classificatory purposes that might sever a number of aims, from explanation and prediction to more practical forms of manipulation and management.

My approach to categorization is functional, on the assumption that in general its purposes can be achieved in a number of different ways, whether alternative or conjoined. For instance, the design and implementation of clustering techniques and the assignation of fuzzy set membership grades are rife with contextual, pragmatic and subjective elements.[1] Understanding categorization admits of multiple perspectives, e.g., properties, groups of causally connected individuals (homeostatic or historical) and a minimal nominalist focus on community-based conventional labels and sortings (including constructed kinds), whether relying on a description (or exclusive definition), a standard (or stereotype) or both, or the exercise of recognitional capacities, etc.[2] While in ordinary life we might favor labels tracking individuals, in scientific representation and methodology, kinds (under any interpretation), populations and properties are primary, even when in different contexts they are conjoined and coordinated into individual concepts and labels that represent and track individuals of interest.[3]

From a linguistic standpoint, we express categorization through kind terms and predication, where predicates too have extrinsic content, the target object of description. We can associate the same predicate 'red', for instance, with different symbolic labels in different symbolic systems such as languages—'rouge'—and graphic styles—'RED'. Typically we take predicates to denote objects in their extension and names to denote their particular bearers, particular red objects.[4] Indexical, or demonstrative, terms are also important in denoting particulars, e.g., 'that red object'. Through an individual, denotation can apply to its class or kind distributively, through its relations of other individuals. How they do so is a matter of controversy. For the sake of sufficient generality, I adopt the minimalism of a methodological nominalist position about signs and individuals, without a nominalist commitment to the exclusion of additional elements.[5]

Since the focus of this book is the role of categorization in understanding and treating vagueness, we should ask whether categorization plays a role in how

---

[1]Cat [1].

[2]For a comprehensive discussion of varieties of approaches to conceptualization and attempts at a common framework, see, for instance, Machéry [2].

[3]A note on language: as a result of my focus on a generalized empirical notion of categorization that includes properties as well as activities, I often speak of categorizations where the reader might expect a reference to categories or even predicates.

[4]Goodman [3], Elgin [4].

[5]I call the strategy, methodological nominalism; as a form of methodological minimalism, it ensures the methodological generality that performs the connecting function my argument requires across a number of conceptual divides.

nouns and indexicals—demonstratives—perform their putative functions and whether in doing so they exhibit vagueness. These are difficult questions that depend on how we understand how these linguistic elements work. But they are also important by analogy with the use of pictures and the interpretation of perceptions. From an epistemic and pragmatic context of communication, whether we succeed in denoting is determined by contextual cues. The same applies to the use of demonstratives and ostensive gestures. From my perspective here, it makes sense to believe that uncertainty accompanies how they pick out something uniquely.[6] Does categorization play a role in such linguistic uses? I think so, whenever it is an empirical fact that the application and understanding of names and indexicals are mediated by kind terms and predicate-based descriptions. In the case of indexicals, we can understand them as abbreviations of descriptions applied to—coordinated with—particular items. For instance, the demonstrative 'that' can be used as short for 'that X' and the indexical 'here' can be used to apply a particular location, that is, a spatial description. Or in the cases of fictional characters, historical figures or individuals and places we are not directly acquainted with, the denotation is mediated by a remembered and shared collection of descriptions or simply reduced to them. All these cases involve categorization and categorization may involve vagueness.

As linguistic symbols, predicates and labels in general denote at least by convention. One may introduce additional considerations of how the convention is applied successfully, for instance, within a context, a community, a system of rules, a level of competence, relative to a standard, a causal connection, etc.[7] Predicates, then, symbolically denote something as having a property or being an instance of a kind or a member of a class, etc. For the sake of generality, as in the case of categorization, I will not express a commitment to a single specific option. Besides allowing for different accounts, one virtue of aiming at generality is accommodating both the possible complexity of multi-factorial combinations and variability across contexts in which different factors or sets thereof could be relevant.

Now, a key dimension of linguistic representation is this: a resulting well-formed predicative proposition is endowed with truth value. The meaning of predicates and the truth value of propositions may be precise or determinate, and thus intelligible, while their intended or unintended correspondence may be said to remain inaccurate. Like other beliefs about our linguistic practices, semantic beliefs that propositions (or beliefs) track the world seem fundamental to communication and action. Understanding how is another issue; whatever truth might be, it is, like belief, a matter of our place in the world that has received multiple theoretical interpretations. Only one of them postulates a transcendent objectivist correspondence with a fixed fact.

---

[6]We can always rule out vagueness by definition, for instance, by adopting standards of rigid designation, in Kripke's sense, relating naming and necessity within a possible-world semantics.

[7]One example of a complex approach is the community-based, multi-component vector model of reference in Putnam [5].

One may note also that mathematical truths by construction and other conceptual truths differ in nature, as indicated by the role that proof or computation play in ascertaining their value. The role of categorization is also correspondingly different. And yet, the constructive dimension of their truth-content, I call it intrinsic content, is analogous to the role relations between ordinary and scientific predicates play in categorizing and establishing matters of fact. In the sciences especially, our so-called knowledge is built and revised on semantic, theoretical and empirical assumptions—they play the scaffolding role of relatively intrinsic content.

2.2 Pictorial Categorization

In symbolic written form, linguistic and mathematical expressions have a visual graphic presentation that we don't consider to be like images in perception or pictures, not even in the geometry of graphs. Writing is not like drawing, at least it is not just drawing, and vice versa. We ordinarily identify differences between a text and figures illustrating the text and also the cognitive difference the figures might make (more on this below). Despite the graphic overlap, they differ as systems of representation and contexts of interpretation and use.[8] In symbolic writing, symbolic design is important but it is not unique, nor it is everything; much more lies in the manipulation and combination of conventional typographic units.

Depiction raises different issues in relation to the roles of perception and categorization. In the pictorial case representation and its uses may rely on precise categorization, but its accuracy is hardly a straightforward matter of truth in the linguistic sense. This is in part due to the fact that the act and relation of representation are not reducible to a single act or relation of predication. For pictures the familiar semantic relation between symbolic representations and their objects is often reversed. Their pictorial role is enabled by being themselves objects of perception and categorization; the perceptual dimension that makes it possible is what I call the *intrinsic content*. One particular version of this cognitive mode of representation based on intrinsic content is Goodman's notion of exemplification of a feature: exemplifying a feature involves reference to it and its instantiation (which may take place through different mechanisms by discerning different kinds of relations).[9]

Regardless of how categorization is established in specific situations, I will assume that shared categorization, *co-categorization*, may be understood in terms of a shared structure—which in turn may be understood in terms of co-instantiation. The categorization involved in representing and recognizing features as something other than the picture itself, I call the *extrinsic content*. When the representation and the represented share properties or categorization, one often speaks of veridicality

---

[8]The overlap includes the existence of ideographic alphabets.
[9]Goodman [3].

or, for perceptual properties, transparency, e.g., an accurate picture of a red rose is red, with the extrinsic content including the attribution of, or reference to, redness; the extrinsic attributive content will include additional categories of flower, plant, etc.

The distinction between *intrinsic* (IC) and *extrinsic* contents (EC) is contextual and their relation also complex through a variety of background assumptions and practices encapsulated in what I call *IC–EC rules*. These rules are associations that often receive symbolic expression and play a key interpretive role also in the formal and empirical sciences (see below for details).[10] An ordinary example is the learned interpretation of visual signals, e.g., road signs. When properties in the extrinsic contents are recognized or declared instantiated by a specific object, they constitute what I call the *external content*. In pictures—the kinds of iconic signs Goodman distinguished by their pictorial, not symbolic, mode of denotation, we may represent in this way increasingly complex and abstract kinds of contents. Representing —that is, constructing and interpreting representations—can be modeled as a process of application of IC–EC rules that enable new levels of exemplification, with new instances of perceived resemblance or otherwise recognized features. The added level of recognition or perceived resemblance still depends on co-categorization. Still, this is no automatic process; the role of IC–EC links is not detached from contexts of their application.

According to some models of perception, our situation in the environment triggers cortical bottom-up processes that involve signals tracking or processing different kinds of robust information from local features of the environment such as orientation, size, color difference or motion (robust modulating properties behave as units of information).[11] Then, binding of those local features takes place into higher-order, more complex patterns of neural activity, such as categories of objects or more complex properties, constrained by top-down selection effects on more basic features. The constraints are set by a so-called tuning of receptors all in a functional relation to memory states, searches, purposes and the possibility or intention of action.

From such a standpoint, a functional notion of perceptual categorization is implemented by functional patterns of neural activation, networks of associations with different capacities involving features of prior experiences and help sort out, predict, or act upon new ones. Memory and perception track classes of experiences and objects in ways that in terms of categorization involve what I call functional resemblance, the sharing of a functional pattern of features. At a conceptual level, we can track those functional categorization events efficiently, by associating with them symbols that enable abstract and general thinking.

With a focus on the role of categorizing activities, a general approach suggests preserving the more functional and empirical aspects, without any commitment to

---

[10]Given the relation between content and categorization, one may read IC-EC rules equivalently, as linking intrinsic and extrinsic categorizations.

[11]In relation to the role of perception in image design, see Ware [6]; see also Gong et al. [7].

specific explanatory entities and mechanisms—neurological, anatomical, social, environmental, etc. Then, also for the sake of generality, one can identify a plurality of uses or purposes associated with categorization—memory recall, classification, inference, prediction, problem-solving, explanation, etc. One should also acknowledge a plurality of types of categorization effective for the job in a given context of conditions—particular or general prototypes or standards, theoretical definitions constraining the application of a variable, social conventions, physical operations, correlations to establish indices and reference, etc. Different types might not be competing models; rather, they might serve the same specific purpose, just differently or in different contexts, or else work jointly—ex., when the application of a criterion involves prior considerations of similarity–; and, vice versa, the same type might serve different purposes. Neuroscientific models such as the one above might be compatible with others, and also help integrate or explain them.

Categorization precedes representation (except in situations where categorization is identified with representation or exercises of a representational function). In the exercise of recognitional capacities appropriately constrained, the activity of categorization also precedes belief.[12] By the same token, categorization precedes similarity. As Goodman already noted, recognition of similarity is ultimately a matter of convention; the interpretation of signs takes place within a system of pictorial and symbolic conventions.

Recognition extends the categorization of the marks that make up a picture into what I call extrinsic content. The activity exerts a cognitive faculty, mainly but not exclusively visual (as other modal cues may contribute to the categorization process), including the imaginative capacity of pretense or make-believe.[13] But it doesn't always rely on an explicit consideration of similarity with the external system and its properties, the external content. This is the domain of perception of images that represent pictorially, not the symbolic domain of the role of perception as in reading.

In fact, as matter of evaluating pictorial representation, different properties of the picture and its putative external target object may trigger the same recognition; that is then the only shared property or respect of perceived similarity. Beyond that, similarity becomes a matter of invariance, correspondence or shared—but not perceived—structure. For the purpose of representation, similarity generally follows the different ways of categorization. We can distinguish between representation and its accuracy; only the latter is in general explicitly concerned with the external content, its knowledge or perception. The picture or symbolic system may represent

---

[12]On the independence of perceptual recognition from belief see Schier [8] and Lopes [9]. Human perception, like machines, might run on a recognition process that follows something like a rule-based iterative algorithm, but if it does, the human algorithm is opaque (Zadeh's terminology). The IC–EC rule is here a matter of mechanism exercising the capacity that enables the formation of extrinsic categorization stimulated by intrinsic categorization. The transparent machine algorithm models the categorization outcome and along the way postulates a procedure that models also the process.

[13]Schier [8], Walton [10], Lopes [9].

what we recognize in terms of the categorizations we generate, but considerations of similarity with the target system is a matter of understanding its accuracy and transparency.

In the case of transparency, seeing is interpreted richly as involving not just seeing-in or seeing-as, but also seeing-through. The debate over this feature of pictures has added similarity to a number of conditions of perception and production of images, for instance, a causal link (indexicality) through natural and intentional modalities of counterfactual dependence, classified accordingly different kinds of pictures and the different modes of their production.[14]

My generalized perspective aims to connect and examine philosophical and scientific objective models, although in a way that does not fix in advance in any context what establishes the content of words and pictures through categorization. They may, at least, have meanings of the different kinds distinguished by Peirce: pictorial (analogical), indexical (causal) and symbolic (conventional). In particular, what I call extrinsic content may yet receive additional interpretation and use in each context in relation to additional background information, skills, purposes, values and standards. The processes and procedures that fix IC–EC links are many.

Vague categorization, then, follows suit as more fundamental than vague representation and partial truth. Objective representation in the linguistic and visual domains represents properties as well as concepts, products as well as processes; for vagueness, if we assume it is a matter of realism as objective representation, what this framework provides is not reduced to misrepresentation, partial representation or overrepresentation (see below); it is meta-representation.

Each specific account of how a given cognitive function is performed in categorization will, in turn, provide an account of the failure of its application in vague instances: for instance, visual blur as a cognitive failure to discriminate, or identify a contrast, between a category and its negation, whether as a basic localized feature or a higher-order binding of local features such as contours, as a result of the overlap of different partially activated neural networks or interference effects between them.

2.3 Mathematical Categorization

In mathematics, set theory provides a basic formal representation for collections of items in general. As a matter of formal application, the theory is developed through the application of a number of rules, concepts and formal techniques; in other words, the representation is constructed in a formal context of application. Whether or not one finds the formal structures rooted in empirical intuition, they have an empirical context of application in which they represent, for instance, the extension of linguistic predicates, e.g., 'red' associated with a collection of (all) red things. In philosophy, the application has long been adopted as a model of concepts and

---

[14]On this debate see Walton [10], Currie [11], Kulvicki [12].

categorization. To categorize is to classify; kinds or the properties their members possess qua members are groupings. One ensuing philosophical debate addresses the question whether a grouping identifies the respective property or kind for those individuals or rather, the property is the criterion that determines the grouping.

Fuzzy set theory provides a generalization of classical set theory. My emphasis on categorization as a practice is not only motivated by how categorization *simpliciter* can accommodate fuzziness as modeled in set theory. Since its inception, one central aim has been to model human categorization behavior expressed by linguistic predicates. This feature is key to its generality; it constitutes a generalization in at least three connected ways: It extends the application of formal set theory to empirical modeling and technological control; it extends the domain of categorizations; and it extends the representation of a concept to a practice of its application linguistically and cognitively, since to apply the formal concept of fuzziness is to represent the practice of categorization and description. Extending the set-theoretic treatment from a model of categorization to a model of reasoning requires in addition a specification of operations that can model linguistic connectives such as 'and' and 'or' and logical constants defined in their terms. The latter will provide models of valid rules of inference.

These practices involve contextual rule-based assessments of similarity to a particular standard; the standard is associated with the recognition of the property in a system in terms of its full membership in the corresponding set and the judgment has its specific validity within the context set by the standard, rules and more pragmatic and subjective factors.[15] The membership judgment is *centered* and *comparative*. The objectivity of the procedures and the assigned feature are related and *relative*, or *relational*.

Now, what is distinctive of fuzzy sets is the assumption that deviations from full membership represent less than full possession of a quality by an individual, or an individual's exact value of another property. The comparative judgment that sets the membership degree is based on a dual standard set by two prototypes, one instantiating full possession of a property, the other instantiating its absence.

The membership judgment is *contextual* because so is the choice of positive and negative prototypes. For the same reason it is also *dynamic*. This feature is neglected in treatments of fuzzy sets. Non-formally speaking, set dynamics has two main sources. The first is the external variability in the cognitive and practical conditions that determine the choice of the prototypes. The second is the shift in the center of gravity of the cluster of cases under consideration as more cases are considered; the extended class may in turn contribute to the change of prototypes in the first source. The contextual membership structure is then holistic and dynamic.

The dual scheme based on two extremes provides, as boiling and freezing points do for the case of temperature, a scale of membership measure. This facilitates the application of particular values to determining the degree of membership or

---

[15]For a discussion of the complexity of this formal practice of set-theoretic representation, see Cat [13].

categorization of a particular case. As a result, the extended range of membership values between 0 and 1 to the real interval [0, 1] can capture the distinction between sharp and fuzzy concepts, categories or predicates; also between exactness and precision, or inaccuracy and vagueness. It is a model of the practice of categorization associated with vague predicates.[16] Precise and accurate values of height, ex., being precisely 5.8 feet in height, or income, ex., earning precisely $100,000, may turn out imprecise at two different levels. One is the degree of membership associated with each precise value;[17] this challenges the notion that a person is precisely 5.8 feet tall. The other is how each precise value qualifies as determined by a related predicate, label, category or property (again, I leave it open which interpretation, linguistic, cognitive or ontic, one may adopt in any given context). Even if the person is may be exactly attributed such precise values of height and income, it is still imprecise whether they are tall or wealthy. Similar examples could be provided in terms of other familiar categories, ordinary and scientific, such as health, safety, and so on.[18]

2.4 Objective Categorization

Vagueness is neither objective nor subjective, neither a feature of real property nor a cognitive state. Since I focus on vagueness of categorization, to introduce my generalized view I need to discuss objective categorization. If the pictorial case carries any objective content or practice, it rests on the complexity of perception. Perception itself is in important cases a matter of pictorial depiction, especially in scientific and technological contexts. Even when higher levels of categorization take place and representational content and power are enhanced as a result of EC–IC rules, perceptual categorization may be required in establishing a minimum set of intrinsic properties and it may likely be rooted in recognition and perceived resemblance. In both such cases, the empirical objectivity of the activity and the associated properties takes a form that is distinctive relational. Perceptual properties such as color or apparent size may be both relational and objective, for instance, as causal interactions between a cognitive system and its environment—after the requisite distinction has been drawn. Then categorization may be objectified with an emphasis on either part in the interaction: as a (relational) property of an object—or system, event, fact, etc.—in an environment or as a (relational) property of the cognitive agent.

---

[16]Zadeh [14].

[17]This is a straight application of Zadeh's formal extension principle and subsequent generalizations; they are rules for generalizing domains of set-theoretic structures and reasoning based on the notion a variable having precise value must be replaced with that of a variable having a degree of membership to each possible value; Dubois and Prade [15], 36–38. Zadeh's original formulation is based on a definition of Cartesian product for fuzzy subsets of different classic universes; see Zadeh [16].

[18]For a discussion of the role of fuzzy concepts in scientific models see Cat [17].

As a property of the system fixing the external content of a picture, its objectivity rests on its material status of its source. As a property of the cognitive agent, it may be considered objectively factual by virtue of a pattern of brain activity part of a process linked to tasks and purposes, with the capacity for external public expression or detection.

I suggest a generalized dual approach to modeling the objectivity of categorization in representation: as a cognitive practice—the subjective or epistemic interpretation—and as its ontic counterpart—the semantic objective interpretation. Ordinary linguistic practices and the construction and application of fuzzy set structures accommodate both interpretations: (1) cognitive categorization is a functional cognitive process, the activity or task of representing or conceptualizing or recognizing, all typically relative to an internal and environmental context; it is also the product of such activities, that is, a representation, with its extrinsic content providing the intensional *precision* conditions for their linguistic expression in predicates and predication, the semantic relation. And (2), ontic categorization is the semantic content; the putatively autonomous object or fact that that stands in a semantic relation and is actually represented instantiates properties individually or in a way extended over groupings. Such so-called independent properties are relational in two ways, as defined over extensions of predicates—e.g., forming a class or a network of family resemblances—and as inseparable from the subjective pole of a cognitive relation, e.g., perceptual properties. The ontic, extensional content I call external content; it provides the *accuracy* conditions for the contents of categories in (1).

(1) and (2) instantiate what I call *sobjectivity*, the inseparability of objective and subjective parts of cognitive the interaction across a set boundary that identifies agent and environment; in (1) it describes the "internal" cognitive process, in (2) its ontic external content. The dual notion can be understood as part of a centered relation of orientation in which the subjective and the objective constitute inseparable terms. The idea can be illustrated with a couple of images: the duality and inseparability is illustrated by the image of the poles of a magnet; the additional oriented-ness is illustrated by the image of a simple telescope, with an appropriately named ocular lens on one end (subject's centered viewpoint) and an a equally named objective or objective lens on the other (object-oriented).

(1) and (2) also distinguish two kinds of models for the language encoding categorization: (1) includes internal representations and patterns of behavior; (2) includes Tarski's disquotational semantics based on truth content, e.g., 'the cat is on the mat' is true of the cat on the mat only if the cat is on the mat; with the classical assumption of semantic determinacy, that is, the uniqueness of intended model. The conceptual issue for the objective set-theoretic understanding of vagueness is whether a scientific theory of truth is a semantic, objectivist notion of truth. The issue is important in relation to the ontic dimension of objectivity and cognition and to the relation of truth of linguistic statements to the accuracy of pictures in relation to their putative content.

To each interpretation of categorization, (1) and (2), corresponds a notion of empirical objectivity; and they are related. The objective pole in type (1) derives its empirical objectivity from the reliability of methodological, rule-based, standards for processes of interaction with an empirical domain.[19] Type (2) includes the possibility of an empirical domain of entities, states or phenomena amenable to empirical interaction. The constraints on the empirical interaction and our evaluation of it relate it to type (1). At the same time, the empirical representation of type (1) is an instance of type (2), as an ontically objective practice, subjective and subject of objective categorization—whether formally, as in fuzzy set theory, or not.

Needless to say, from a nominalist standpoint, (2) is just a form of (1). From an empiricist standpoint, both constitute kinds of accessible factual reality in the world, whether as properties of things or as cognitive habits, activities, functions and processes—social or neural. In type (1), categories and representations bear syntactic properties that are themselves categorized and instantiated, especially perceptual ones in pictures. On my interpretation, fuzzy set theory is an account of categorization that invites the objective inseparability of cognitive and ontic modes of objectivity.

At the same time, as I have noted, we have to acknowledge the contextual character of the assignation of specific degrees of membership in fuzzy sets; it is the cost of the mechanically formal, methodological objectivity that characterizes (1) and, on some interpretation, yields (2).[20] But then, the way (1) informs (2) implies an important qualification to the kind of ontic objectivity in objectivist interpretations of vagueness (see next chapter).[21] As a product of (1), the fuzzy form of (2) represents indeed an objective feature, but just as a measurement result might; it is not just any objective feature. The category defines a relational property determined by the structure and prototypes distinctive of the context that gives meaning to the objective membership degree values associated with the objects categorized.

The application of fuzzy set theory suggests another qualification. Subjective elements—affective, perspectival, normative and volitional—are part and parcel of the context and the practice of assigning numerical membership values. Elsewhere I have discussed how the role subjectivity plays in formal fuzzy categorization practices involves both ontic and epistemic aspects. Minimally, I take subjectivity to include singular individual perspective, situated cognitively and embodied physically, and expressed in the exercise of skills, judgment and other activities.

But do they undermine the outcome's ontic objective interpretation? We can distinguish between two basic notions of objectivity, content (product-centered) and methodological (process-centered) objectivity. Subjective elements certainly limit any *methodological objectivity* that rests on formal procedures or the systematic application of any rules. In particular, the procedures involved fail to yield a numerical outcome in a univocal, determinate manner. We may think of this failure

---

[19]Other things being equal, standards include calibration and replication.
[20]Cat [13].
[21]See, for instance, Smith [18].

of univocality as a sort of *practical vagueness*. But once the outcome has been otherwise generated, the decision been made, the judgment been issued, its *content objectivity* becomes a residual matter of ontological commitments to ascertain how the numerical magnitude might correspond to any ontically objective empirical reality. From a realist perspective, the degree of membership may be precise yet simply inaccurate.

Vague categorization tracks, then, vague representation ontically and cognitively, as representation of vagueness in the world of the object represented and the world of the visual representation. Representation often rests on an element of inference; that's how categorization is often extended, below I call this content development, and IC–EC rules play a role; and vice versa, inference builds on representation. What follows is that the cognitive interpretation of the process of categorization and ontic interpretation of the categorization process and outcome are inseparable.

Then, there is a role for the broader notion of objectivity I claim is at play in vagueness and its formal fuzzy modeling. Independently of the difference between both objective kinds of vagueness in their role in representation, vagueness can be categorized as a manifest property of representations, regardless of its either epistemic or objective interpretations. Either way, fuzzy membership, without partaking of a semantics of partial truth, can still model that kind of vagueness in the world and categorization representing it. But it suggests a different approach to its interpretation and application in the domain of representation and reasoning with images.

References

1. Cat, J. (2016). The performative construction of natural kinds: Mathematical application as practice. In C. Kendig (Ed.), *Natural kinds and classification in scientific practice* (pp. 87–105). Abingdon: Routledge.
2. Machéry, E. (2009). *Doing without concepts*. New York: Oxford University Press.
3. Goodman, N. (1976). *Languages of art*. Indianapolis: Hackett.
4. Elgin, C. Z. (1997). *Between the absolute and the arbitrary*. Ithaca, NY: Cornell University Press.
5. Putnam, H. (1978). Realism and reason. In *Meaning and the moral sciences* (pp. 125–126). London: Routledge.
6. Ware, C. (2008). *Visual thinking for design*. New York: Morgan Kaufmann.
7. Gong, S., McKenna, S. J., & Psarrou, A. (2000). *Dynamic vision*. London: Imperial College Press.
8. Schier, F. (1986). *Deeper into pictures*. Cambridge: Cambridge University Press.
9. Lopes, D. (1996). *Understanding pictures*. Oxford: Oxford University Press.
10. Walton, K. (1990). *Mimesis as make-believe: On the foundations of the representational arts*. Cambridge, MA: Harvard University Press.
11. Currie, G. (1995). *Image and mind: Film, philosophy and cognitive science*. New York: Cambridge University Press.
12. Kulvicki, J. (2014). *Images*. New York: Routledge.

13. Cat, J. (2015). An informal meditation on empiricism and approximation in fuzzy logic and fuzzy set theory: Between subjectivity and normativity. In R. Seising, E. Trillas, & J. Kacprzyk (Eds.), *Fuzzy logic: Towards the future* (pp. 179–234). Berlin: Springer.
14. Zadeh, L. A. (1965). Fuzzy sets. *Information and Control, 201*, 240–256.
15. Dubois, D., & Prade, H. (1980). *Fuzzy sets and systems. Theory and applications.* New York: Academic Press.
16. Zadeh, L. A. (1975). The concept of a linguistic variable and its application to approximate reasoning. *Information Science 8*, 199–249, 301–357; *9*, 43–80.
17. Cat, J. (2006). On fuzzy empiricism and fuzzy-set models of causality: What is all the fuzz about? *Philosophy of Science, 73*(1), 26–41.
18. Smith, N. J. J. (2010). *Vagueness and degrees of truth.* Oxford: Oxford University Press.

Chapter 3
Vagueness and Fuzziness in Words and Predication

In philosophical debate, how to best understand the vagueness of predicates remains a controversial matter. A renewed defense of epistemic interpretations, in terms of deficient states of knowledge with insufficient information or a semantic relation of indeterminacy of interpretation (whether as content or predicate), has been met with a critical reaction and prompted a defense of vagueness' objective character.[1] On objectivist views such as the one Smith has recently defended, vagueness is not a semantic relation, but an ontic feature of the things denoted by the subject of predication. The feature is then expressed by a semantic relation of partial truth to the predicative proposition, in accordance with fuzzy set theory.[2]

In this chapter I borrow conceptual resources deployed in Smith's defense of objective vagueness to test the limitations of its focus on the linguistic model when applied in the case of pictures; this will allow me to explore and exploit the common framework behind the application of fuzzy sets to both words and digital images. If fuzziness is a model of vagueness, I want to explore whether the same criterion of vagueness applies to pictorial representation and how the conclusion yields new insights into the empirical application of the mathematical concept of fuzziness.

To that effect I adopt a generalized version of the linguistic standard of vague representation that relies on the semantics of degrees of truth provided by fuzzy set theory. I already introduced in the previous chapter the senses in which categorization and fuzzy sets modeling it can objectively describe both kinds of facts, qualities in the world as well as practices and judgments of categorization, which also take place in the world. As a matter of our categorization practices, vagueness might be appropriately recognized in cases of epistemic nature (for epistemic or cognitive reasons). My broader, empirically objective approach to categorization, truth and representation seeks to accommodate just this kind of possibility. The ensuing conceptual standard of

---

[1]Williamson [1], Keefe [2].
[2]Smith [3].

© Springer International Publishing AG 2017
J. Cat, *Fuzzy Pictures as Philosophical Problem and Scientific Practice*,
Studies in Fuzziness and Soft Computing 348,
DOI 10.1007/978-3-319-47190-7_3

vague categorization provides the bridge criterion for extending the concept to pictorial representation. But this move requires that we focus on the role of categorization, with all the caveats mentioned before; along the way I will show how the pictorial case differs and so does the corresponding interpretation and application of fuzzy models.

Vagueness in predicates may be associated with a number of indicators of limited relevance, especially the adjudication of a borderline characteristic of sharp concepts with decidable conditions of application. Smith distinguishes two problems in vagueness: the problem of determining the boundary of predicates (*location problem*), and the problem of determining whether they are either sharp or blurred (*jolt problem*). The second is where vagueness becomes a potential matter of indeterminacy.

Standard criteria of vagueness are related and neither ranks more fundamental than another.[3] In order to keep track of a systematic comparison with the pictorial case, I use the following notation, V(l) for the criteria applied to linguistic case of predicates and V(p) for the corresponding application in the case of pictures. The criteria are following: (1) $V_1(l)$, vague predicates give rise to borderline cases; (2) $V_2(l)$, vague predicates have extensions with blurry boundaries; and (3) $V_3(l)$, they generate Sorites paradoxes in deductive arguments.

To defend the relevance of fuzziness as a model of objective vagueness, Smith adopts another criterion, $V_4(l)$, the closeness condition, as more fundamental and a solution to the jolt problem: A predicate F is vague just in case for any two objects they are close in respects relevant to F; in turn, closeness in respect of F is characterized semantically by a necessary condition in terms of truth: statements of each object being F are very close in respect of truth. If F is vague, then, for any two objects, statements that they are F are close in respect to truth.[4]

Amount of height and degree of error represent degrees of accuracy in the application of the predicate 'height'; by contrast, degree of tallness is a matter of degree of precision, and its application claims a matter of degrees of truth. What is true of truth values is true of the models of a theory or language; truth in intermediate degrees correspond to properties in intermediate degrees. For the content of predicates in those propositions, then, Smith can conclude that vagueness is in the world, not in the semantic relation of indeterminacy.

This is the relation that, according to him, warrants the fuzzy ontic interpretation of vagueness; the degree of closeness is measured by a degree of truth based on degrees of membership in sets associated with predicates in question.

I want to suggest that the closeness criterion as well as the borderline can be modeled formally more generally by extending fuzzy sets with properties from so-called rough and near sets. In rough sets members are characterized by indistinguishability, the matching or identity of description values. In near sets members are characterized by tolerance, the closeness of description values below a set amount, including matches (more in Part 2, below). Tolerance of difference-making

---

[3]Here I follow Smith.
[4]Smith [3], Chap. 3.

becomes a matter of closeness in truth; but in pictorial cases, the closeness of content must be first in respect of accuracy, that is, what may qualify as pictorial accuracy or fit.

Now we can contrast the putative failures of two dimensions of determinacy, uniqueness and univocality. Whenever determinacy of a predicate or a statement containing it obtains, a model can decidably make a statement either true or false; and, on the ontic assumption attached to the semantic relation, the feature symbolized by the predicate is instantiated in the model. What may be called overdeterminacy, or, better yet, plurideterminacy, involves the assumption of a fact of the matter about the truth, about there being something to be true about, but it's just not determined uniquely by the facts or their description (depending on whether a model is understood as a truth-making entity or a predicate's contents). In such situations, a predicate or a statement containing it may have different models or interpretations; conversely, something is a model of multiple representations or contents, e.g., a subject is capable of multiple predicates or a fact multiple descriptions. We may consider uniqueness in terms of a robust overlap of adequate univocal interpretations or determinations (this is the model of precision in plurivaluationist accounts of vagueness).

The semantic interest in models of predicative propositions has led to a dual notion of a *model*: as a truth-maker for a statement and as a particular type of representation of phenomena, entities, states of affairs, etc. The type of representation has been interpreted as an abstract structure that constitutes a truth-maker for hypotheses about it. Of course, scale models and prototypes are yet additional notions involving similarity relations. Pictures can be understood as models in at least the representational sense, as visual representations or *visualizations* of models—where models are understood as more abstract representations that may received multiple expressions or concrete representations either pictorial or linguistic. Still, my approach to categorization emphasizes the role of pictures also as models in the first sense, as objects of description alongside their own intended objects.

In the case of indeterminacy, there is no fact of the matter about truth or correct interpretation. Determinacy lacks the univocality of uniqueness. Not even one model, whether as intensional or extensional content, meets the predicate. The determination is at best a mysterious partial affair. While ambiguity, for instance, is a form of plurideterminacy or even pluri-indeterminacy (multiple-model indeterminacy), vagueness may be considered associated at least with a form of single-model indeterminacy. Not every potentially relevant object is a clear model of a sentence with the relevant predicate, or for a given object, this doesn't qualify clearly as a model of a specific predicate or statement in its terms, or else an instance of a specific category. In objectivist views such as Smith's, vagueness is a kind of indeterminacy only in appearance; what seems like a challenging semantic relationship is objectively a determinate matter of a degree of truth, a partial model, a feature of the property instantiated in the world. In my generalized view, vagueness may be in context judged to be an epistemic matter; and truth in the objective semantic sense may be also a matter of cognitive facts about predication and its indeterminacy.

Vagueness may not be reduced to single-model indeterminacy. But semantic indeterminacy appears besides: The adequate or acceptable semantic content of a sentence, the proposition and its truth-making interpretation, may still not be unique. This is the result of the limitations of our semantic capacity of discrimination, the resolution of our meaning-fixing practices, independently of vagueness —e.g., classical plurivaluationism. This is, according to Smith, a (typically epistemic) problem of determination of the boundary of predicates (location problem), regardless of the question of their nature, sharp or blurred (jolt problem).

Fuzzy categorization, as I noted above, was originally introduced precisely as an empirical mathematical model of phenomena such as vagueness in the use of linguistic predicates to represent categorizations, in other words, a model of a cognitive activity and linguistic practice. Smith's analysis makes clear how the model can be both justified and interpreted; although here I argue that both can be extended. The empirical application of the formal structure of a fuzzy set isn't simple; as many clustering techniques, it often relies on rules and judgments of similarity to others instances of the category in question, especially one adopted in a given context as the standard of degree 1 of membership, and often also the other end of the range, one case of membership degree zero.[5] For their cognitive role a number of different criteria of similarity and measures of similarity degrees have been proposed.[6]

In quantitative systematics the challenges of articulating and applying a category of classification are no closer to the ideal. As one author put it, 'numerical phonetics has not succeeded in developing a method of producing a single relatively objective classification, better than others, which could be used as the formal classification of any group.' And, he added, 'the selection of characters, the coefficients of similarity or difference to be used, and the methods of clustering are all determined subjectively, and there are no generally accepted criteria for deciding which of the various classifications is 'best.'[7] As in the case of fuzzy membership functions, also for the basic taxonomic methods a number of different criteria of similarity and measures of similarity degrees have been proposed. The difficulty has suggested also a fuzzy-set version of higher-order vagueness of borderlines, the so-called type-2 fuzziness in which degrees of membership are themselves fuzzy.[8] Higher-order

---

[5]Another possibility is that an ideal prototype is not in the category or (non-normalized) set defining it, or that multiple and incommensurable prototypes; Dubois and Prade [4]. Conceptually and quantitatively, the fuzzy set-theoretic approach offers a generalization of the qualitative empirical models of graded categorization in research by Eleanor Rosch and others in the 1970s, and in fact it also precedes them.

[6]See Klir and Yuan [5] for examples of set-theoretic criteria, typically in terms of set operations such as intersection and of metric quantities, or distances, defined over feature space with each coordinates referring to the value of the degree of membership in a set/feature, and as coordinates as dimensions or features.

[7]Michener [6].

[8]The fuzziness of fuzziness can be represented in type-m sets with m indefinitely large.

fuzziness is an expression of the difficulty, cognitive and methodological, assigning degrees of membership in practical situations.[9] Of course, this property complicates the ontic objectivist interpretation of vagueness in terms of truth and properties. The degrees of truth are themselves vague, so it the possession or instantiation of a property. The picture becomes more complex.

Also degrees of belief may play a role in practice.[10] Considerations of similarity, respects, categorization and predication reappear, as we shall see, in accounts of visual representation. Even any degrees of belief associated with vagueness may be said to correspond to objective strengths of tendencies to act; this feature might be said to extend similarly to the action-ready content of visual representation, first in our perception.

From an empirically objective standpoint, fuzzy theory provides as many objective models of vagueness as there are proposals for how to estimate membership functions and assign values, or at least how to constrain their estimation. The procedures loosely resemble procedures for quantitative measurement or translation; and they share an ambiguous status that admits of a dual interpretation, objective/subjective and ontic/epistemic.

The following is a selection of examples of different types of procedures illustrating their overall diversity.[11] (1) Discrete exemplification: this procedure fixes a set of hedging terms that qualify for truth values ranging from true to false; each is then exemplified numerically by a corresponding membership value so that each truth term is translated into a membership value term ready to be applied to the relevant cases. (2) Deformable prototypes: a concrete prototype is modeled by a set of parameters; then a dissimilarity distance measures the minimal parameter difference between prototype and object after each parameter has been manipulated, or "deformed", with a required amount of energy of distortion; the membership function of the maximally matching object is the complement of the normalized energy-weighted dissimilarity function. (3) Implicit analytical definition: the marginal increase in the strength of the belief that a predicate applies to an object, 'x is A', is proportional to the strength of 'x is A' and the strength of 'x is not A'. (4) Likelihood: the increase in membership value is proportional to the increase in the probability of a positive response by a randomly chosen individual to a poll question about categorizing x as A, where x might also be the polled individual herself (a more constrained version is known as the voting model). (6) Statistical measure: membership expresses the proportion of positive responses. (7) Risk or bet: membership measures the level of risk associated with a statement, expressed in the form of a corresponding bet; as a measure of commitment or trust it is said to express subjective probabilities. (8) Relative preferences: membership values of elements in a set are derivative, calculated from relative values with respect to another element's in the form of absolute value ratios. (9) Subset comparison: the

---

[9]Dubois and Prade [4], 30.

[10]See Klir and Yuan [5], Dubois and Prade [4], Cat [7].

[11]Dubois and Prade [4] and Lawry [8].

average membership value of a subset of elements determines whether a it matches the original better than another does provided its average is higher. (10) Filter function: the membership function is a linear function, the filter function, distributed symmetrically over a population segment centered around a so-called neutral point 0.5, so that outside the interval, objects get full membership, 1, or non-membership, 0.

Similar considerations apply to related proposals to measure the degree of fuzziness of a set: 'the difficulty of deciding which elements belong and which do not belong in a set.'[12] Decision difficulties constitute a sort of volitional indeterminacy that is also present in semantic practices of language users; especially in connection with the jolt problem and boundary cases, where speakers have difficulties deciding whether a predicate applies. As such, these sorts of difficulties cut across the product-process, ontic-epistemic and objective-subjective divides. We reencounter the same ambiguity in the type of objective contents of fuzziness representations.

There are two main types of measures of degree of set fuzziness: distance and entropy. Distance measures are aggregated differences of membership values for set elements with membership within a range. One simple proposal for entropy function is defined by analogy with the Shannon entropy function in classical information theory. In the information-theoretic case, the function measures the average uncertainty involving the prediction of outcomes in random experiments and is defined for probability distributions over members of a set. In the fuzzy case the function is similarly defined over distributions of membership values of a set.[13] The analogy to the objective measure of informational uncertainty combines the ontic interpretation of membership and the cognitive dimension in terms of volitional uncertainty. Information itself, much like fuzziness, is used and interpreted in explanatory models on the basis of my generalized, dual notion of objectivity.

The additional dimension of objectivity is constituted by the status of fuzziness of categorization as the object of empirical and formal investigation. As I pointed out in Chap. 1, above, the subject/object distinction, like organism/environment and other dichotomies, is not arbitrary, but a practically effective moment in an interaction; whether one adopts an ontic interpretation beyond cognitive—empirical and methodological—standards of objectivity and truth depends on making additional objectivist assumptions about truth, "reality" and the properties of objects—entities, facts, sates of affairs, structures, events, etc.

References

1. Williamson, T. (1994). *Vagueness*. London: Routledge.
2. Keefe, R. (2000). *Theories of vagueness*. Cambridge: Cambridge University Press.
3. Smith, N. J. J. (2010). *Vagueness and degrees of truth*. Oxford: Oxford University Press.

---

[12]Dubois and Prade [4].
[13]Ibid., 33 and Klir and Yuan [5], Chap. 9.

4. Dubois, D., & Prade, H. (1980). *Fuzzy sets and systems. Theory and applications.* New York: Academic Press.
5. Klir, G. J., & Yuan, B. (1995). *Fuzzy sets and fuzzy logic. Theory and applications.* Upper Saddle River, NJ: Prentice Hall.
6. Michener, C. D. (1970). Diverse approaches to systematics', *Evolutionary Biology* 4 (1970). In G. Dunn, B.S. Everitt (Eds.), *An introduction to mathematical taxonomy.* Cambridge: Cambridge University Press, 1982.
7. Cat, J. (2015). An informal meditation on empiricism and approximation in fuzzy logic and fuzzy set theory: Between subjectivity and normativity. In R. Seising, E. Trillas, & J. Kacprzyk (Eds.), *Fuzzy logic: Towards the future* (pp. 179–234). Berlin: Springer.
8. Lawry, J. (2006). *Modelling and reasoning with vague concepts.* New York: Springer.

Chapter 4
Representations: From Words to Images

We have a set of criteria; next, we must test their extended applicability. This requires examining the roles of perception and categorization, and relating them. Instead of aiming to suppress any differences from the conceptual vagueness of linguistic predicates, I intend to bring them out and exploit them. After pointing to a few distinctive features, I examine representation in visual perception and then continue on to the perception of pictures.

We can find characteristics that differentiate images as distinct kinds of representations. For instance, formal, uninterpreted, so-called syntactic features of pictures, may bear a distinctive arrangement and density of marks that distinguish analogical from digital images, and, correspondingly, pictorial from linguistic (or symbolic) systems.[1] This is the domain of features that make up the intrinsic content (IC), always distinguished in context. It is a proper subject of categorization in its own right, represented and recognized as instantiating perceptual properties—and, by that standard, also representing them. For this reason I focus on visible pictorial representation in (mostly) analogical pictures.

If the intrinsic level can be the locus of vague categorization, it must assume a sufficiently rich and diverse role for perception, especially seeing and their related modalities, seeing-in, seeing-as and seeing-through (see below). Again, my analysis should be detailed and developed further than I do here in light of specific empirical findings—properly interpreted. Seeing something in the marks in a picture or in some object's features points to the intrinsic pole of the experience, while seeing them as the object in question points to the extrinsic pole. At the expense of seeing the medium or awareness of the process of the picture's production, the experience of seeing the object in the extrinsic content is often reported as the experience of transparency or seeing-through. This is the standard of perceptual realism; throughout the post-medieval history of Western imaging practices it has been associated with ordinary conditions of perception and has been interpreted as an

---

[1]Goodman [1].

© Springer International Publishing AG 2017

J. Cat, *Fuzzy Pictures as Philosophical Problem and Scientific Practice*,
Studies in Fuzziness and Soft Computing 348,
DOI 10.1007/978-3-319-47190-7_4

instance of direct realism or involving some kind of mediating mental representation—a kind of realist representationalism (see next chapter).

This variety of cognitive modes and roles associated with perception constitutes the source of intrinsic and, indirectly, extrinsic contents; each provides a potential source of vague and indeterminate categorization. Intrinsic visibility is a distinctive feature that has no significant counterpart in language, in predicates precise or vague, beyond, of course, the visual (or audible) recognition of the token linguistic signs bearing symbolic meaning, by convention.

Pictures are two-dimensional spatial arrangements of meaningful visual features; symbols have arbitrary visual features but form meaningful linear or serial arrangements. Each kind belongs in systems with different kinds of features and rules of interpretation and use. Some combinatorial kinds of diagrams have features of linguistic symbols. The role of spatial relations and other visual properties of a picture are key to the representation of similar properties of its external content or referent. Arrangements of visible, spatial properties such as structures or patterns are responsible for its capacity to denote pictorially.[2]

Moreover, pictures often satisfy the Fodor–Sober semantic compositionality condition: all parts of a picture depict a corresponding parts of the system identified as the picture's target or object.[3] Since the condition doesn't fix how many parts or level of composition are required, it is compatible with typical holistic aspects of pictorial depiction.[4] Recognition and representation emerge at some of the possible levels. There exists a limitation from below. Semantic compositionality in pictorial representation relation can be expressed by the degree of precision or definition of images. The size of the smallest representing part sets the level of resolution. One extreme is the one set by Goodman's density criterion. According to it, the analogical base size is infinitesimally small; pictures depict or predicate pictorially to the extent that they are analogically dense; the semantic condition applies to every point as the smallest discernible part. This feature explains the comparative richness of potential visual information over linguistic symbols.

But then this means that images might not depict pictorially according to Goodman's limiting standard. Since early in the history of photography, the chemical and optical molecularity of film was identified as setting the limit to the medium's visual precision, extolled as already surpassing that of painting. In digital images, the degree of definition is set by the size of its smallest discrete pictorial unit, the pixel. In a molecular medium the precision of an analogical picture, of its visible tonal and geometric features, is related to how perceptual properties are described—that is, categorized—of the medium's effective pictorial units.

Again, the challenge of identifying the possibility of vagueness, from the standpoint of the linguistic precedent and the applicability of fuzzy-set formal

---

[2]Pictorial denotation involves analogical spatial density of signs (syntactic) and signified (semantic) Goodman [1].

[3]Sober [2], Rollins [3].

[4]Schier [4].

models, lies with the role of categorization. I have pointed to one dimension of pictures where the issue arises. This is where accounts of pictorial representation such as Hopkins' identify the distinctive nature of pictures: for a base class of things that a picture represents as occupied, or, equivalently, a category instantiated, a linguistic description represents a genus as occupied by instances of a species (the associated propositional content with truth value), but not vice versa (the class isn't a genus for some species of members).[5] Pictorial depiction fixes the species of the genus represented as occupied (it cannot represented color without representing a determined color). And vice versa, it might be thought to represent, inferentially, every genus covering the species represented as occupied (if it represents a color, it also represents color; if a colored animal, also animality, etc.), but it doesn't fix it. By contrast, linguistic description doesn't fix the species by fixing the genus; but it can fix the genus. Pictures are less generic, more specific or fine-grained in terms of determined, fixed content involving more levels of determinate content (e.g., species as an occupied base class).

More features have been noted for images in general within the so-called non-conceptual content debate about mental representations: Images are characterized in terms not reducible to the conceptual content of linguistic representation. Linguistic symbols contribute coarser-grained representation with abstract information in horizontal or same-level articulation of potential conceptual contents; images present, by contrast, finer-grained representations (similar to analogical density) with a vertical, cross-level articulation of potential conceptual content (it cannot represented color without representing a determined color), and with image parts representing parts of the target system, although the relevant image parts, in terms of neuronal distributions, are not always spatial, nor are they indefinitely small.[6]

All these features are relevant to the possibility of visual thinking (see next chapter). Can images then be responsible for categorization? From the point of view of prototype theories, networks or clusters of perceptions associated with the object of categorization, e.g., individual systems, events, situations, phenomena, or data points, would play the role of categorization. We can then distinguish between two situations in which words and images relate differently: in one case linguistic labels provide the symbolic, even computational, encoding of categories; in the other, images merely illustrate the conceptual relations encoded symbolically. From a phenomenological viewpoint, the two situations are often indistinguishable. The distinction, as I note next chapter in relation to thinking more generally, is either a sematic or an explanatory issue. To the extent that categorization is a form of thinking and part of others, defenders of the role of embodied and visual cognition endorse the irreducibility, if not the primacy, of the non-symbolic, non-computational representation and activity we associate with imaging.

---

[5]Hopkins [5].
[6]Rollins [3], Kosslyn [6], Kulvicki [7].

As a matter of communication, it is obvious that words and images interact in effective ways; pictures might be said to be worth a thousand words, but those thousand may not be sufficient in order to add sufficient determination to the intended content. Titles and captions, symbolic notations and words on maps and graphs, narrative voice-over in film, all suggest that linguistic expressions do not report on content of images, rendering them redundant. Although with linguistic descriptions we can attribute to them a more determinate content that also render the corresponding statements more or less correct or incorrect. They act in an auxiliary capacity regardless of their intrinsic vagueness. And vice versa, images may provide templates for linguistic or symbolic communication, e.g., maps. The relation of word to image is a complex and contextual affair. A similar difference and complicated auxiliary relation obtains between simpler and sharper diagrams and rich and fuzzier photographic images so often juxtaposed or superimposed for clearer categorization and analysis (see next chapter).[7]

The richness of visual representations is however richness of relevant and recognizable or categorizable content. This is a perceptual standard of realism; pictures provide an opportunity for perceptual searches and information gathering similar to the equally time-consuming practice of scanning our visible environment (albeit within a more limited range of bodily motion). Perceptual realism is potential and dynamical realism. It is this dimension that makes photographs, for instance, such effective enduring substitutes for first-hand fleeting perceptual situations. If we try to elucidate the cognitive value of images, the different dimensions of thinking with them and not just from them, they prove themselves reliable tools for exploration and analysis besides sources of evidence.[8]

If photographs and other pictures may be thought of as useful form of visual prosthetics, we must recall that their successful operation requires familiarity with the conventions, aims and capacities they modulate and skill handling them.[9] There is one recent digital replica of the dynamical aspect of our searching searching capacity, computational field photography, also known as plenoptic photography. A dense array of micro-lenses enables the camera to record light sources within a broad range of distances and angles, then software allows the viewer to select and explore the resulting digital environment by refocusing on different angles and distances.

Despite the noted differences, the features of picture-making or imaging processes do not set visual representations radically apart from linguistic expressions. On the one hand, linguistic (and mathematical) symbols themselves are visible (except in exclusively oral traditions); that visibility is key to dominant practices of recording and communication of information. On the other, the contexts in which systems of visual representation are crafted, transmitted, applied and appreciated

---

[7]Lynch [8].

[8]On the evidential role, see Perini [14], Meskin and Cohen [10].

[9]On the prosthetic approach to photography see Lopes [11]; on the complexity and contingency of conditions of depiction see Lopes [12].

reveal their conventional dimension. Their uses do not it a dichotomy between conventional symbols and natural signs imitating the world.

Inference and representation in linguistic statements are based on the semantic relation of truth; also does objective vagueness. Do pictures tell the truth? Is truth-bearing a condition for performing their cognitive functions? The linguistic standard suggests the possibility of visual truth conditions, but whether they are available is a controversial matter. In the linguistic sense, we may attribute truth values to propositions and other truth-functional intentional states such as beliefs; and in certain conditions we may associate those with pictures and their perception. But this fact does not make the pictures true or false. They may just be related to perceptual beliefs in such away that the extrinsic content of pictorial representation is adopted as the content of a truth valued proposition or belief; then we say the picture becomes the source of evidence.[10] Yet the picture is still not true.

From the fact that we judge the representational function of pictures we may infer that we consider them able to be accurate or veridical, that they correspond or match, and they do so to different degrees of approximation. What can truth be? On one proposal of an extended notion, for instance, truth within symbol systems may be formulated within extensions of Tarski's recursive semantic model such as Perini's.[11] Elements in such an approach are characters naming by virtue of the formal properties and interpretive conventions within the system of symbols and linguistic expressions. Cited features in the classical symbolic model are: discrete syntactic structure and univocal determinacy of reference within symbol systems, recursive compositionality of symbols for statements and their truth value, and rules for translation of names for statements into higher-order meta-languages. Truth, in an extended semantics, may take the form of a correlation in meta-language between a linguistic name and a linguistic expression.

In this type of account, the reliance on reference secures the possibility of realism familiar in accounts of linguistic symbols defending realism even without representation.[12] By parity of reasoning, the decoupling of representation from realism suggests the possibility of representation without realism; this might be, for instance, the case of intrinsic content and of extrinsic content without accuracy. Also my approach to categorization, above, decouples objectivity from realism. If we think of representation in terms of the categorization that fixes the intrinsic and extrinsic contents, these and other considerations, below, suggest limiting the validity of the assumption that might called the *intrinsic-extrinsic content link*, that intrinsic content fixes extrinsic content.

Alongside the suggestions that emphasize weaker forms of semantic relations such as correspondence without truth, one may consider the application of epistemic standards of correctness, that, as in the case of a great deal of beliefs and

---

[10]A version of this possibility is suggested in Walden [13].

[11]Perini [9].

[12]Recall Kripke and Putnam's causal account of reference and its role in Putnam's defense of realism about theoretical entities across paradigm shifts between incommensurable descriptions.

hypotheses, rely on grounds of valid inferences or reliable mechanisms. Attempts to label this general alternative have introduced a considerable number of terms such as 'correspondence', 'conformation', 'veridicality', 'correctness', 'match', 'fit', 'isomorphism', etc. This is weaker than the case of the phenomenon of recognition involving categorization without judgments of resemblance between the picture and its object. However, my approach differs from the universalist or reductive formulation of many specific proposals. As in the case of the forms of categorization and the possible functions of images, also representation I think is a matter of local and plural criteria.

The classical model is more readily applicable to visual symbols. Diagrams can be given a semantics in terms of a system with a discrete and articulate syntax for well-formed figures (and rules for forming them), a semantics (specifying denotation) and assigned truth values. Graphs can have parts named and given linguistic translation, that is, their content represented linguistically.

But can analogical pictorial representations bear truth? The issue is, why truth matters and whether weaker alternatives to recursive semantic definitions for symbol systems are available. When engaging pictures, their valued roles revolve around representation, communication, exploration and reasoning. Truth-bearing or any equivalent semantic property might not be a requirement on pictures such as diagrams for discharging those cognitive functions in all cases.[13] For instance, chemical structural diagrams may act as names denoting symbolically or models denoting descriptively so that claims about them may be made as they are made about models and inferences may be made about the corresponding properties of represented compounds.[14] In the absence of truth or a semantic substitute, vagueness and the exercise of cognitive functions that manifest it such as representation are better considered at least a matter of categorization.

References

1. Goodman, N. (1976). *Languages of art*. Indianapolis: Hackett.
2. Sober, E. (1976). Mental representations. *Synthese, 33*, 101–148.
3. Rollins, M. (1989). *Mental imagery: On the limits of cognitive science*. New Haven: Yale University Press.
4. Schier, F. (1986). *Deeper into pictures*. Cambridge: Cambridge University Press.
5. Hopkins, Ch. (1998). *Picture, image, and experience*. Cambridge: Cambridge University Press.
6. Kosslyn, S. M. (1994). *Image and brain*. Cambridge, MA: MIT Press.
7. Kulvicki, J. (2014). *Images*. New York: Routledge.
8. Lynch, M. (1985). Visibility. *Social Studies of Science, 15*(1), 37–66.

---

[13]Goodwin [15].

[14]Goodwin attributes each function to different but compatible representational approaches, Goodman's about signs and Giere's about models. See below.

9. Perini, L. (2005). Visual representations and confirmation. *Philosophy of Science, 72,* 913–926.
10. Meskin, A., & Cohen, J. (2008). Photographs as evidence. In S. Walden (Ed.), *Photography and philosophy. Essays on the pencil of nature* (pp. 70–90). New York: Routledge.
11. Lopes, D. (2008). True appreciation. In S. Walden (Ed.), *Photography and philosophy. Essays on the pencil of nature* (pp. 210–230). New York: Routledge.
12. Lopes, D. (1996). *Understanding pictures.* Oxford: Oxford University Press.
13. Walden, S. (2008). Truth in photography. In S. Walden (Ed.), *Photography and philosophy. Essays on the pencil of nature* (pp. 91–110). New York: Routledge.
14. Perini, L. (2005). The truth in pictures. *Philosophy of Science, 72,* 262–285.
15. Goodwin, W. (2009). Visual representation in science. *Philosophy of Science, 76,* 372–390.

Chapter 5
Epistemology, Aesthetics and Pragmatics of Scientific and Other Images: Visualization, Representation and Reasoning

Vagueness of appearance and depiction is a property of the categorization of images. Through categorization, whatever pictures do, they may do approximately and vaguely. And what images can do, that is, what they do for us and we can do with them, depends on what we think their roles are. In general, pictures play a role in ordinary and scientific argument and in cognition more broadly. They are key to identifying, documenting, tracking and exploring visible properties of empirical systems and phenomena, also and to visualizing and communicating empirical and theoretical information; they can be emotionally compelling, aesthetically powerful, and exhibit and enforce values and biases. This is no less relevant in the study of systems, individual or generic, whose relevant properties are spatial, chromatic or structural.[1] Relevant examples differ widely in medium, mode of production and use; they include photographs, drawings, data charts, diagrams, animations, film recordings, computer generated images, etc. Pictures in many such cases are meant to support inferences, recognition processes and carry heuristic and evidence value. We think with them and through them.

In science visual thinking takes place in a number of related ways involving visualized conceptions, ideals or theoretical hypotheses as well as visual and visualized data.[2] I call *pictorial empiricism* this cognitive extension of empiricism, understood even grounded in experimental settings and practices, from observations[3] of relevant features of systems of interest (often called target systems) to observations of artificially produced visual data. We may say that pictorial empiricism is part of intervention-driven experimental empiricism. From a

---

[1]The literature is vast. In the scientific cases, see, for instance, Baigrie [1], Larkin and Simon [2], Tufte [43], Daston and Galison [3], Hentschel [4], Perini [5], Goodwin [6] and Kulvicki [7]. Some of the literature focuses of generalities or taxonomies; another draws attention to the contextual and contingent nature of particular cases.

[2]Empirical methodology often works on the basis of contextual discrimination of (relatively) theoretical stuff. The distinction may be based on different criteria.

[3]Arguably, relevant experience that in scientific inquiry plays the empirical role of perceptual input may not be strictly visual.

© Springer International Publishing AG 2017
J. Cat, *Fuzzy Pictures as Philosophical Problem and Scientific Practice*,
Studies in Fuzziness and Soft Computing 348,
DOI 10.1007/978-3-319-47190-7_5

perceptual standpoint, a further extension incorporates the visualizations, visual graphic illustrations of abstract ideas with prior symbolic expression. Both visual and visualized thinking, then, are activities relevant to pictorial empiricism.

I want to draw cursory attention to a number of forms and functions of pictures in scientific practice and in the rest of our cognitive life for a simple reason: However we make them and whatever we can do with them, categorization will allow for the possibility of making them vague and using them vaguely.

In science, the selective, constructive and representational aspects of pictures are familiar tools, as are their cognitive uses in computation, explanation, etc. These are all features of modeling practices and are characterize, by construction, models of phenomena or entities, whether real or fictional. Whatever models may be, like fictions, we represent them by representing a selection of features; they are those we decide are particularly salient within a particular context for a particular purpose and from a particular standpoint.

In fact, a number of accounts of what models are and how they represent have been inspired by accounts in aesthetics of what images and fictions are and how pictures represent—and much else—whether by means of similarity, indexical denotation, information, projection (illusion) or pretense.[4]

Many pictures are effectively *visual models*, and some of them are *visualizations* of models. Diagrammatic elements juxtaposed to or superimposed on dense figurative photographic images make this point clear. The photograph stands in for the complex reality—or its perception— that calls for parsing and categorization in order to acquire relevant cognitive and practical value.[5]

As I noted earlier, the notion of model has acquired in logic and philosophy of science a diversity of meanings; I have emphasized two: models as truth-makers or content of symbolic statements and models as representations. Each is a term in a different semantic relation of correspondence, fit or accuracy. In science, the symbolic statements are associated with formal expressions of theories (syntactic structure). Those sorts of statements are the ones whose representational contents is denoted conventionally and they are made true in a direct fashion, e.g., Tarski-style, by representing something that is in fact, that is the case. In quantitative terms, mathematical formulation of the hypotheses and mathematical measurement of the empirical features allow us to express and assess the standard of ideal correspondence in terms of the intermediary criterion, e.g., quantitative approximation. In the last section and in more detail elsewhere, I note that this is in itself a complex matter.[6]

Because models are constructed and adopted also as representations—for the purpose of understanding, explanation, etc.—, they do not represent their target systems in the same way. The bearers of truth and numerical accuracy are

---

[4]Frigg and Hartmann [8], Giere [9], Bailer-Jones [10], Weisberg [11].
[5]Lynch [12].
[6]Cat [13].

associated linguistic hypotheses about them.[7] This is the sense in which models are like images and images operate as visual models. Also in the case of science, denotation is linked with relations of forms of Goodmanian co-instantiation. The philosophical relation of fit is often described as a relation of isomorphism, invariance of structures or similarity between the model and its target system, visible or invisible, real or fictional. In an abstract sense of similarity, they all exemplify different, more specific forms of similarity.

I will not pursue this issue further; my concern is how similarity is only one way in which categorization and representation are related in visual depiction and I examine the issue further in the next sections as a potential source of pictorial vagueness. The case of scientific images is no more intriguing than the accounts that defend and explain models and the importance of modeling.[8] It adds to the different ways in which, in science and elsewhere, visualization and modeling contribute another type of vague representation and of anything else that relies on categorization.

Representation, reasoning and much else in science may take a visual form and a potentially vague one too. It is important to note here that representation and reasoning, or thinking more generally, are interdependent practices; just as reasoning depends on representations, representing may involve reasoning.

To what I want to draw passing attention in scientific cases is not the notion that similarity may hold a mysterious and controversial philosophical place. Instead, I want to point to its role as a standard of representation and reasoning adopted in science practice.

Analogy has long been recognized and studied as a heuristic and a mode of inductive reasoning.[9] The inductive aspect is expressed as probabilistic uncertainty or incompleteness; both are considered forms of qualitative approximation, and are assessed in terms of a difference in the numbers of features. It is another question whether analogical thinking is in fact a genuine form of reasoning for the purpose of establishing valid inferences or a helpful set of rules with heuristic value to suggest hypotheses. I believe that as an inductive kind of argument, the calculus of similarities is based on judgments that bear different possible interpretations but may be used in support of uncertain conclusions. Cognitive psychologists have examined its pervasive and complex use in ordinary cognition.[10]

As a tool for scientific representation, similarity has received formal attention and empirical application. We can distinguish two main related criteria of similarity: metric and set-theoretic. Metric criteria are geometrical analogs in the form of measures of proximity. They play a central role in mathematical theories of categorization and taxonomy. Whether simple matrices, matching coefficients or complex clustering algorithms, these various kinds of measures are defined

---

[7]Giere [9].
[8]Kulvicki [7].
[9]Hesse [14].
[10]Gentner et al. [15], Hofstadter and Sander [16].

set-theoretically, namely, over discrete or continuous sets of possible values, or states, for characters forming discrete or continuous sets.[11] Similarity, or dissimilarity, relations are restricted relations of equivalence: reflexive, symmetrical and, in a restricted number of cases, transitive. Then, if the relevant conditions obtain, similarities form equivalence classes; they form disjoint sharp sets.[12] On this basis distance functions can be introduced as measures of dissimilarity satisfying triangular inequalities, just like elements of length in different geometries. The base set-theoretic relations can take complex forms of intersection relations, as in the case of so-called near sets and tolerance relations, where equivalence classes are introduced between different sets of grounds of relations of descriptive indistinguishability or identity.[13]

Similarity relations are not only rooted in categorization but they help establish categorization. How they do so and when is a complex contextual matter. From the qualitative standpoint, in relation to the set of features or characters different systems may share, formal measures provide degrees of similarity or fit separate from relation of numerical accuracy.[14] In the last section, below, I refer to them as instances of conceptual approximation, and dimensions also of visual approximation.

Now, if similarity is based on categorization and is in turn an instrument for categorization, then it is another possible source of vagueness, beyond degree of similarity in features or degree of accuracy in value, and is amenable to the formal treatment in terms of fuzziness in representation and reasoning, and in the case of perceptual similarities, visual representation and reasoning. Fuzzy-set generalizations are based on particular fuzzy-set generalizations at least of membership functions (state values) and the relation of transitivity.[15]

Again, among the roles of depiction categorization supports, representation and reasoning are inseparable. Evidence or inference is not entirely separable from representation, as one can specify notions of representation that will inform inference and vice versa. For instance, categorization often relies on inference, as does recognition of similarity; and vice versa, inference relies on categorization. If categorizations can exhibit this relation, so does any interpretation of images based on them. One sort of content, intrinsic, may provide the basis for inferring another, extrinsic, or develop the latter further. As a matter of evidence and truth, as I

---

[11]Dunn and Everitt [17].

[12]The ordinary psychology of analogical judgments presents asymmetries and other contextual features first detected by Tverski; this, as well as other chronological and cognitive factors, may be linked to the asymmetric function of analogy for instance in the generation and interpretation of metaphors, also in science; see Gentner et al. [15], Cat [18].

[13]Set similarity is introduced to define near sets in Peters and Pal [19], 1.7. Set similarity ultimately rests on measures of indistinguishability of points relative to features defined on them, that is, the difference between values of feature probe functions, or membership functions.

[14]For an application of set-theoretic criterion in the case of the relation between models and their target, see Weisberg [11].

[15]Dubois and Prade [20], Klir and Yuan [21].

mentioned above, the process involves causal relations to images, beliefs and hypotheses that bear truth and evidential value relative to other propositions. But in my view, the relation to empirical reality that helps establish reliability in argument is the relation that supports categorization. On those grounds we make decisions as well as inferences, and, as a result, we are able to solve conceptual and practical problems. And we can do so in a public manner.

This is the problem of empirical inquiry, and the role of pictures is an extension of the role of individual personal perception. Thanks to socially and materially objective dimensions, then, pictures can play the same role as observations and visual data and extend their use within a broader social context of virtual witnessing and evaluation. They support perception, communication and thinking beyond the symbolic record of private observation events or the access to specific technologies of physical interaction with our environment, whether analogical observation and symbolic detection. More importantly, they do so in ways that help manage symbolic information, conceptual or numerical, or replace it altogether.

A case in point is simulation. Simulations visualize and animate numerical computations, allowing scientists to process the information in productive and efficient ways.[16] Another case is big data. Even big data sets may be modeled through formal treatments and also visualized to form significant and informative patterns. Besides an issue of effective communication of intended numerical or conceptual information, their visual recognition is common in identifying phenomena or signs of relevant phenomena in stock market analysis, medicine or particle physics alike. Visualization might help categorize meaningfully and make diagnostic inferences efficiently, for explanation, prediction or intervention. It can also play a role in evidentiary reasoning, for the purpose of testing or confirmation; for example, the tracking of visual patterns of data can provides the basis for assessing the qualitative fit between the visualized set of data and the form of a model's predictions. From reasoning to recognition, these are all cases of relevant categorization, whether the outcome of transparent mechanical procedures or of subjective and opaque processes. The different kinds of categorizations involved will provide a source of vagueness in addition to any inaccuracy or approximation, qualitative or quantitative. I discuss the notion of pictorial approximation in Part 2, below.

In formal and natural languages and reasoning, general rules establish the methods of inference about particulars and the generalizations. Reasoning, it is claimed, rests on truth[17] and generality (abstraction).[18] Symbolization is an extreme form of abstraction for the sake of accuracy;[19] it is based on conventional rules of formal semantics replacing intuitive (i.e., pictorial) similarity or abstraction. From the set-theoretic formal models of vagueness in natural languages, fuzzy theory has

---

[16]It is common to confuse animation and simulation mistaking one for the other.
[17]Perini [5].
[18]Shin [22].
[19]Shin [22].

provided a general framework for a number of proposals of systems of fuzzy reasoning.[20]

By formalizing units of an image or by using the image as a source of observational statements and inferences from them that we might make up a linguistic argument in support of a hypothesis. This is the point of empiricism. But there's no systematic mapping that exhausts the relevant information and its generative power. In the case of dense images, additional evaluative criteria of relevance for shifts in optical density across contiguous pictorial units are required. One needs perceptual premises that correlate the visible features of the image with those of the alleged content. But this cannot explain how images support determinations of their content, that is, corresponding beliefs about the system they depict (the role of judgment is pervasive in cases of adopting or applying clustering criteria) or how they support general or explanatory hypotheses. Nor can transcriptions or reconstructions eliminate images and perceptual information from the different cognitive processes. As I have discussed above, pictorial and symbolic depictions interact in complex ways and may help add determinateness to each other's content, but they don't stack up neatly so that their content may be interchangeable and one representation replaceable. This situation has consequences for the interpretation of fuzzy set theory I discuss in Part 2, below.

If reasoning is a matter of evaluating and establishing the validity of inferences and extending our stock of reliable information, thinking may be a wider-ranging matter, involving other cognitive tasks.

An alternative to thinking with formal reasoning about true linguistic statements is thinking in the context of communication, with the aim to help induce belief. It is the broader domain of rhetoric through the role of figures of speech such as the familiar analogy, metaphor, hyperbole and synecdoche. Their vagueness of meaning plays an important role in communication. The effect and use of rhetorical strategies when we speak figuratively is not formalizable in terms of truth and logical rules, as when we speak literally. Statements containing rhetorical devices such as metaphors—'Juliet is the Sun'—are strictly speaking false. Yet the play an important role in the formulation and communication of scientific claims.[21]

Now, this kind of cognitive intervention extends to visual communication. Pictures play an equally fuzzy figurative role in visual argument through techniques of visual rhetoric. From visual analogies to visual metaphors and caricatures, we recognize some kind of distortion within specific contexts. For the rhetorical interpretation and their role in persuasion to be plausible, also there the matter of meaning is one of categorization, understood by loose analogy with rhetorical types of figurative language. Their role in argument is linked to their more or less fuzzy content. This is important because it enables its place within linguistic discourse that informs the effective context of persuasion. Moreover, in both modes of

---

[20]Klir and Yuan [21].

[21]I discuss views and examples of scientific metaphors in Cat [18].

communication we often associate the figurative functions of categorization with the cognitive value of concrete cases and experiences.

Also cognitive tasks such as retrieving information in memory and constructing representations in imagination may rely on internal visualizations or external pictures. Generally speaking, we use images in varieties of reasoning, computing and problem-solving of different kinds, even if how the brain processes images is a matter of debate about their actual availability and role.

Visual problem-solving relies on spatial thinking.[22] For instance, we use maps of different kinds to navigate the world and follow instructions encoded in images rather than instructions encoded in linguistic sentences.[23] Once we identify our location on a map—by entering the visual fiction of the cartographic model, even conditionally, by analogy—we can determine the relative location of places of interest, their distance between them along a path that, if followed, will allow us to get there. The representation of the terms of the problems and their possible solutions is hardly trivial. I develop this issue in chapter 19.

In pictorial representation or communication, visual recognition has a holistic dimension with semantic and epistemic dimensions. But holism as a matter of interpretation and evidence differs from the molecular or digital limits of the Sober–Fodor condition. Units of interpretation might be available but their meaning or evidentiary value might not be aggregative. Patterns on weather maps or MRI-images of body parts do not show storms or cancer formations that can be visually analyzed into the visual evidentiary units of stormy or cancerous conditions. Even when maps include units of iconic meaning explained in a key, the distribution of the iconic units of information is displayed on a holistic background of geographic formations. The recognition of facial features relative to a knowledge base is a common challenge with multiple applications (especially in relation to the ellusive role of cultural differences).

The aggregative use of the units of economic or geographic information becomes relevantly meaningful against the organizing holistic background of the chart or map as a whole. Symbolic or analogical, typical displays that call for interpretation and provide evidence rely on units of meaning but the content and evidentiary value of the picture as a whole is not reducible to similar value for any given part. It is at best a hybrid case. Recursive and atomic semantics are more readily available in the symbolic meaning of diagrams and their rules of construction and interpretation.

Some of the cognitive role in visual thinking involve the use of visual designs we call diagrams. They are, distinctively, kinds of pictures and not just perceptual images of the environment. Diagrams may have a variety of cognitive virtues and serve a variety of purposes. Graphic visualization of empirical and theoretical hypotheses may serve purposes of computation, evidence or prediction. Think of the computational function of Feynman diagrams in quantum field physics

---

[22]A range of issues related to so-called visuospatial thinking are discussed in Shah and Miyake [23].

[23]Kosslyn [24], Taylor [25].

Fig. 5.1 Aerial photographic image of Chicago with diagrams drawing attention and identifying the presence and location of sailboats. Photograph by the author

and diagrams of velocities, weights and forces in classical mechanics—central even in Newton's geometrical, not analytical, proofs of his famous dynamical explanations in the *Principia*. The semantic, representational content does not always carry descriptive or explanatory value, or carry it exclusively, as in the role of geometrical representations of area and length to visualize proportions. Visualization of empirical data too may serve equally computational purposes, even displaying a similar lack of explanatory relevance, but it might serve predictive purposes: for instance, alongside auxiliary empirical correlations providing the interpretive rules of thumb even without underpinning explanatory theories; think of geometric rules in stock market analysis. Notice that their role is independent from different models that try to make sense of how we carry out tasks beyond basic skills involved in spatial thinking, for instance, comprehending quantitative information in graphical displays.[24]

This is evident in the pervasive circulation of superimposed or juxtaposed pairings of precise diagrams and imprecise photographs (Fig. 5.1).[25]

---

[24]For an integrative and evaluative review of such models see Shah et al. [26].

[25]Lynch [12]. In a forthcoming essay Scott Curtis distinguishes between the aesthetic of the smooth and the rough, with the focus is on what I have been calling intrinsic visual features of the images.

In relation to other images the diagrams plays, in general, two connected mediating roles: it places the analogical dense picture in a series of connected categorizations, a *categorization chain*, and connects the visual representation with symbolic and linguistic representations with prior relevance to cognitive and practical tasks. The diagram isolates and delimits precisely parts of the image, and in doing so it facilitates the identification at a different level of categorization within the relevant conceptual framework of analysis, interest and purpose.

With sharper, simpler linear diagrams we introduce a graphic link between the intrinsic content—the vaguely localized chromatic speck—and the extrinsic categorization—sailboats in Chicago. The key issue is the connection between different levels of relative precision. But the precision can characterize equally what we consider the intrinsic visual features or content of the picture and extrinsic ones. Vagueness or imprecision may be relative to higher-level more extrinsic categorizations answering the question, 'what is that?,' or begin with intrinsic categorization that first defines "that" so that we can ask questions about it. The diagram provides the guiding vehicle through different levels of categorization and their communication.

We typically find closed lines such as circles and open lines such as arrows. They connect by performing two different categorization roles. By intrinsic contrast, sharper, simpler and contrasting geometric and chromatic features, they can draw our attention to a part of the image they point to. The role of cue by learned association involves an indexical and performative sort of categorization: 'Look here,' 'look at this.' 'This' and 'here' are more precisely defined visuospatial categorizations by means of the graphic aids. We recognize the meaning of the symbolic lines, whose design choice is not purely conventional, and we have learned to use them adopting them as successful instruments of communication.

We can also use the open and closed lines representationally in a more direct connection to the more specific or more abstract and general identification we consider relevant. In this kind of categorization, lines bear a more direct analogical relation to the geometric features of the system they mean to "carve out" or extract. We superimpose more complex shapes approximating the theoretically pre- supposed standard of structural features that we want to represent—e.g., the shape of a boat, a building or a tumor, rather than their location— and to link them symbolically to the relevant classification that attributes the entity in question an identity or kind—for some cognitive or practical purpose of prediction, explanation, intervention, etc.). With these diagrams we rely—either by design or interpretation—on visual contiguity by superposition or juxtaposition to attribute to a certain part of the image an external content, representing something "real," specifically a certain structural property with a 2-dimensional or 3-dimensional geometric representation. With arrows we can introduce dynamical representations between image parts in the form of information about motion or causal action. The "reality" aspect is the indexical function of the image that in general I associate with the causal, material conditions of interaction that enable relevant categorization. Typical examples include medical X-ray images, microbiological electron microscopic images, telescopic astronomical images, archeological images and aerial satellite survey images.

The performative and the analogical functions are not completely distinct. The circle itself can operate as a more precise representation of the shape of a certain system of interest. We can connect the basic graphic elements such as the circle with the attention-guiding performative role with the more specific representational lines by deformation.

In the second, mediating, cognitive role, additional lines direct our attention also performatively outside the picture or the relevant isolated part, or by proximity, allowing us to attach the next level of categorization expressed symbolically, either linguistically or quantitatively. Additional information is expressed linguistically in the figure's caption or title and in a longer accompanying text. With all this visual design and information combined, the picture presents and integrates in fact two *content maps*, an *intrinsic map* and an *extrinsic map*, which in the photographic case amount to an *external map*, that is, a categorization map of external target content. In the case of superimposed diagrams and words, both maps overlap in the same spatial region.

Distinguishing the pictorial from the linguistic might prove a helpful heuristic, but the approach to vagueness I explore here includes the value of acknowledging and understanding their similarities and interactions; this is the motivation for the emphasis in categorization and its multiple expressions.

This mediation between the pictorial and the symbolic (aside from symbolic uses of pictures) plays two cognitive roles: illustration and evidence. In both cases it's less a matter of logic than of connection to a privileged concrete kind of representation that may contribute to thinking. Familiarity with the perceptual content of descriptions provides a cognitive connection we value for different purposes, as a cognitive and practical aid. Pictures also extend the domain of the role of more direct observations of things, events, etc. in our empirical world, the world we interact with perceptually. The link to observation and to causal connection to parts of the world channels the factual realism or whatever epistemic status we grant our mediating model of empirical reality to the epistemic status of some hypothetical claim. In doing so the connection of established perceptual beliefs and the hypothetical beliefs provide evidence.

Often the connection requires the transformation of qualitative modeling of the individual experience or its public pictorial record as data framed within the strictures of a code or format of representation. Often, through measurement techniques, the modeling is quantitative; in that case, the evidential support of relevant claims is effected by formal chains of mathematical codification and computation with added rules of inference. This involves complex and diverse methodological practices.

From an epistemic standpoint, how we understand scientific results, information or knowledge will be inseparable from how we understand the kind of support that the codified chains of categorization provide. The causal interaction with "reality" might suggest accepting results on grounds of their revisable reliability, whether descriptive or pragmatic (if separable). Alternatively, it might suggest acceptace on grounds of more audacious commitments to factual truth (conformity, correspondence, etc.) and reification based on additional levels of categorization and hypothesis, especially a commitment to the existence of entities and properties explanatory models pretend to describe accurately.

We can conclude that the categorization chain that provides conceptual insight and empirical support is not the performance of an isolated and linear logical proof. Nor is it an aggregate of interpretations of smaller perceptual units. It is embedded holistically and contextually in a complex cognitive scaffolding that includes layers of information about other parts of the picture and its production as a whole. They include empirical and hypothetical background assumptions and practices that include the appropriate use of technological means of interaction with the environment that includes the picture's content. In Fig. 5.1, the technology includes a plane and a camera and the author's body connecting them. What appears like an isolated narrow visual focus is in fact the eye of a cognitive tornado.

A related source of diagrams' role in representation and reasoning is the role of generality. Representation and recognition of properties are often supported by matching or identifying some similarity to members of a set. In this respect, as I have mentioned above, diagrams combine elements of pictorial and symbolic representation, diagrams combine features of models and notational signs, pictorial and symbolic forms of notation and denotation: one based on pattern recognition and emphasizing the concrete and different, the other based on convention—social and effectively arbitrary- and emphasizing the similar, general and abstract. The range of contributing elements makes them able to present different modes of fuzziness affecting the diagrams different functions.

Reasoning, especially proofs, involves processes of both individuation and generalization; so do diagrams.[26] Here is the generalization problem: how to bridge the conceptual, epistemic or logic gap from particular to general, especially in reasoning. The problem has been solved along two approaches to formalization:[27] (1) Lockean abstraction, by means symbolic formalization with arbitrary symbols in replacement of token properties and semantic stipulations, with inference rules for well-formed formulas, for the sake of conceptual and inferential accuracy (conceptual determination and truth); and (2) Berkeleyan denotation, by means of pictorial formalization with diagrams based on selective attention to token properties, with transformation rules for well-formed token figures, for the sake of pictorial, concrete, intuitiveness.[28] Symbolic accuracy prevents token-dependence by stipulation; diagrammatic accuracy prevents token-dependence by systematic ruled transformation. Diagrams too, not just symbols, are reliable tools for general conceptual representation and reasoning (and not just a cognitive heuristic).

Alongside generalization, there is a role for individuation, that is, the consideration of particulars in categorization and reasoning by construction or by stipulation. In linear symbolic systems, substituting for particulars, rules ensure the elimination of ambiguity. The spatial nature of pictorial diagrams requires the

---

[26]Shin [22].

[27]Ibid.

[28]See Barwise and Etchemendy [27], Barwise and Hammer [28], Shin [22].

idealized representations of particulars, and this enables the creative generation of richer representational structures. These structures are endowed with heuristic value in the form of possible different paths of conception and pursuit of a proof, especially in geometry. Among these different paths are the selection of different auxiliary individuals and the exhibition of information contained in visual structure, otherwise assumed implicit. The latter demonstrates by display what otherwise needs to be inferred.[29]

The practice relies on different resources of varying degrees of generality. For instance, geometric thinking is just one set of diverse analogical, constructive and inferential practices that play a role in visual thinking, including diagrammatic thinking, without exhausting them. A related practice is reasoning with geometric patterns, that is, through the recognition of structures as spatial configurations that may be modeled as structured sets with specific relations between members. They provide visual templates for categorization and analogical reasoning, for instance identifying common patterns of descent and gene distribution. As an empirical mode of reasoning, it relies on empirical background assumptions and semantic criteria in biology—e.g., individuals, reproduction, traits, units of inheritance, etc. Formally, it is an application of a more basic diagrammatic mode of reasoning that plays an important role in mathematic understanding, proof and computation.[30] It relies both on constructive activities of mental visualization—that is, imagination—and the manipulation of external representations. In this way the same resources may be applied to visual reasoning with data, from clustering, ruled-based machine learning and heuristic formation of hypotheses.

Diagrams can model and help carry out otherwise symbolic proofs and can form modes of reasoning that satisfy the requirements imposed on symbolic forms. For instance, Frege developed a notational system of logical diagrams for carrying out proofs in propositional logic.[31] The simplest and best-known case is the use of the diagrammatic tradition with Eulerian and Venn diagrams to illustrate set-theoretic relations and model logical connectives and carry out syllogistic reasoning (Fig. 5.2).

Applying Venn diagrams to the visualization of empirical data adequately categorized allows for simple forms of empirical reasoning. In particular, it allows for simple forms of causal reasoning. Data falling within subsets or partial subsets—that is, co-categorized—suggest correlations and evidence for causal relations either as relevant phenomena or explanatory hypotheses. The conceptual error is to adopt the set-theoretic relation as a causal criterion that identifies such relations as causal.[32]

---

[29]Shin [22]; on Gestalt perception of information in diagrams see Coliva [29].

[30]See Resnik [30], Giaquinto [31].

[31]For a recent defense of Frege's diagrammatic notation see Dirk Schlimm's 'On Frege' Begriffsschrift notation for propositional logic' (Univ. of McGill ms., 2016).

[32]I have argued against such moves in the application of fuzzy set theory in Cat [32].

Fig. 5.2 Venn diagrams
illustrating and guiding a
syllogism *Left* All *A*s are
B. all *C*s are *A*. Therefore, all
*C*s are *B*. *Right* No *A* is *B*. All
C are *B*. Therefore, no *C* is *A*

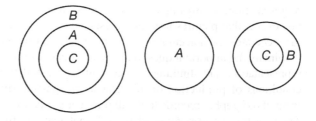

The use of diagrams in reasoning raises the issue of transcription or the cognitive equivalence supported by the mapping between picture-assisted cognitive tasks and rule-based symbolic inference and computation. We can distinguish two relations. In one case, the non-equivalence between linguistic symbols and diagrams is pragmatic, in terms of differences in cognitive benefits and supporting conditions such as skills. Diagrammatized token individuals are better than symbolized individuals as problem-solving and hypothesis-forming heuristics. Even when symbolic rules are available, their generative role facilitating cognitive tasks is not reducible to the application of symbolic expressions.

In the second case, the mapping is not readily available, if at all. At best it is an ad hoc contextual mapping rather than the application of general symbolic rules. Picture-based thinking often relies on empirical constraints and background assumptions that determine the nature of the cognitive tasks, e.g., the analysis of images and the use of pictorial information to make calculations and predictions. This is the problem of selective transcription I have already noted in relation to the possible interactions between pictorial and linguistic descriptions and their respective indeterminacy. In fact, not all cognitive roles are illustrations of rules of inference or computation. In both cases, diagrammatic pictorial formalization is widely considered a uniquely effective cognitive resource in categorization, problem-solving, communication and reasoning.

In cases of standard formal reasoning under specific semantic interpretations, the problem of transcription may depend on introducing logical operators and rules connecting truth values. One strategy is to map pictorial and symbolic truth conditions. But, does visual reasoning depend on visual truth? We are back to the question of need and availability of truth values, or something equivalent for the purpose, such as accuracy and reference, required for inference and use. Are all pictorial representations expressible in a language with a syntax and semantics? While Goodwin denies the requirement, graphs and diagrams might be, and this fact allow for the formulation of rules of inference. Natural languages lack precise symbolic formalization of notions of truth and rules of reasoning. So do analogical pictures. But reference and accuracy of representation might provide the closest to what's needed for a shared linguistic and non-linguistic semantics that can support truth and reasoning with language and pictures. Representation of states of affairs, or reference to them, might be what is available to us, and accuracy of representation; it also might suffice at least in practice.[33] The common assumption however is that pictures used in representation and visual reasoning are precise, not just accurate—or inaccurate—, but that might not be what in practice is available.

Finally the important question is, where does fuzziness come in in tasks of visual representation and thinking? When and how are they vague? The technological conditions of production of the relevant images constitute sources of vagueness, from photographic records to the digital representation of data points and diagrams. Then, in the categorization of the relevant features fuzziness gets its conceptual expression.

Fuzziness may appear in intrinsic visual parameters such as topological properties of boundaries, geometrical properties of shapes or chromatic properties of colors. Analogical images may present intrinsic fuzziness in their role as extensions of perception as a standard of empirical realism; they yield a key type of pictures familiar in visualizations of detection events. Familiar situations include digital face recognition, the digital imagery of spatial distributions of chemical substances or organic matter in water, the distribution of spectral lines in the atomic analysis of light, ultrasound visualizations of organic structures, the X-ray photographs of bone structures, satellite imagery, telescopic observations of geological features of planetary surfaces, or displays of weather conditions.

A more ordinary example is the design and use of maps. Fuzziness of intrinsic features appears in the limits of discrimination of the graphic elements—based on material and other constraints derived from the use of a medium, driving interests and purposes, etc. Lines and dots—small circles—may not capture any location or boundary according to a more precise, discriminating shape.

Not all fuzziness will appear in this analogical way. The second type extends the analogical display of detection of spatial features to the design of visual models of data or hypotheses. Maps, schematic figures and graphs sit in the overlap between the spatially analogical and the diagrammatic. In the case of diagrams, the fuzziness of categorization and the thinking to which it may contribute concerns what in a given context I have called extrinsic interpretive content. In this sense, it is similar to the case of fuzzy categorization of precise quantitative data.

In diagrams the visual expression of fuzzy categorization—content, meaning or information—can take different forms according to the different possible IC–EC links. Fuzziness may characterize the picture's extrinsic content as well as the intrinsic content. When the extrinsic content is vague, the information associated with some element is uncertain, its interpretation is objectively imprecise—from the ontic perspective, in relevant cases it depicts an indeterminate feature of a system or phenomenon–; by contrast the intrinsic visual features may be as precise as we recognize them for the purpose at hand. For instance, part of a medical or financial graph, a kick or the size of a dot representing a data point, may lack precise

[33]Perini [5], Siegel [33].

meaning as a depiction of a system or situation in the terms associated with the dimensions of the graph.[34]

When the intrinsic, visual content is fuzzy, some of its visual, design features are fuzzy—always relative to a baseline standard in a practical, sensory context—in a way often perceptually indistinguishable from seeing them fuzzily, e.g., fuzzy color contrasts, shapes or edges. We can encounter two types of situations, which, again, my form overlapping or continuous groups: the fuzzy visualization may be the outcome of the interaction between the target system and the technological elements of detection and visualization—e.g., an electrocardiogram–; and it may be a design choice, the intended visual expression of the believed extrinsic content. In the simple case of Venn diagrams the fuzziness of the categorization of individuals as members of a set can get two different visual designs: in one the individuals may be depicted with fuzzy-looking small smudges; in the other, each closed curve representing a set is depicted with a fuzzy edge rather than a sharp enough boundary. A quantitative alternative in fuzzy set theory depicts the precise membership grade without fuzzy-looking lines, edges, shapes or colors with, for instance, fuzzy-set Venn-type and hypercube representations (see Part 2).

Now, diagrams can be used to illustrate and carry out deductive reasoning, not just computational inferences; therefore, the kind of vagueness of categorization represented by those diagrams might fall under the pictorial version $V_3(p)$ of criterion $V_3(l)$, namely, the possibility of Sorites paradoxes.[35]

5.1 A Note on Practice-Dense Data Visualization

Content extension takes place by means of a sequence of acts of visual categorization. Whether or not the different layers of interpretation may be ordered into levels, they are the product of a complex, contextual set of practices of manipulation and quasi-theorizing associated with the classification of data—and thereby of theoretical phenomena corresponding to data sets adopted as bodies of evidence. In data-based pictorial empiricism, classification of data, visual or not, is itself a practice of categorization and it involves processes of production, interpretation, selection and representation; the last dimension is where we can find visualization. Any partial formal rules for interpretation of the form EC–EC' only black-box a wealth of activities.

In fact, a regulated set of similar practices applies to visualization itself. Visual marks are data to be categorized and may be visualized in turn as graphic data models expressing further categorization, and so on. That is, data visualization qualifies itself as data in the cognitive functional sense. These cognitive processes are mutually reinforcing and rife with added practices, values, assumptions, etc.

---

[34]This is different from the meaninglessness of the thickness of the graph, or the size of the map.

[35]For a treatment of the linguistic case with fuzzy inference rules, see Trillas and Uturbey [34].

In addition, visual and visualized data may play a role in thinking as well as being an instance thereof. By thinking, here I mean categorization, problem-solving, computation and inference. Graphic visualization, imaging in general, is itself a form of selection based on considerations of reliability and often of evidence.

So, what can data be in general? Data are the epistemically significant elements of observable information. Specifically, they are observable and mobile—portable —marks or objects, traces, signs of interactions with the world, or symbolic expressions or graphic records thereof, informed by different kinds of techniques, devices and conceptual (even theoretical), symbolic and experimental perspectives, skills and manipulations.

For authors such as Leonelli, Bogen and Woodward, data are public traces or records, whether marks, photographs, graphs, numbers;[36] yet data aren't transparent representations of known properties of phenomena. Instead, they argue, we should adopt a relational, functional, epistemic account of data: 'any object can be considered as a datum as long as (1) it is treated as potential evidence for one or more claims about phenomena, and (2) it is possible to circulate it among individuals'.[37] The evidential role of data is based on the explanatory role of phenomena causing the observable public records. But for the explanatory role depends on the operations of the particular instruments that produce the data in particular contexts and conditions, material and social.

The question, what are data?, leads us to another, what are data for? A methodological, indeed epistemic dimension emerges early informing the process of their production, and this leads to the promise of their epistemic use value. The two are inseparable. And epistemic roles abound: Evidence for hypotheses about entities, processes and phenomena (Bogen and Woodward, Leonelli); prediction; and pattern discovery and theorizing (Leonelli's classificatory theory) for the sake of further using, interpreting and collecting data sets, old and new, that are associated with phenomena, entities or processes. Many such roles often involved the pictorial expression and perception of the relevant information.

The task of classification or categorization of data shows the complex and contextual character of practices involved. In other words, no formal rule can do justice to them as a ready-made substitute. Formally and practically convenient as this may be, an uncritical and simplistic approach to the context of formalization and automation poses methodological and practical risks.

No automated inferential procedures applied to data, such as induction, have general significance and reliability. Instead, valuable discrimination and critical understanding require acknowledging a process based inter alia on diverse kinds of background knowledge, social interactions, model-based reasoning and iterative manipulations of materials and instruments (know-how, with or without

---

[36]Bogen and Woodward [35], Woodward [36], Leonelli [37]. See, for instance, also Cat [38] and Hentschel [4] for the case of photography.
[37]Leonelli [37].

corresponding metadata, its background knowledge).[38] This complexity is evident in how practices of description and experimentation are interdependent. Meaningful classification requires acquaintance with processes of production and collection of the data and all sorts of constraints imposed on them, from material conditions to methodological standards and research interests.[39] As a result, formulating found associations in terms of rules for extending content can be only of limited reliability and generality, yielding at best somewhat stable expressions of local, contextual complex processes.

Bio-ontologies illustrate rich forms, methodological conditions and cognitive roles of categorization. They are networks of general and generalizable terms, defined by descriptions and sets of empirical evidence.[40] Biological classification in networks of bio-ontology terms sorts phenomena, entities and processes, that is, the ontology, through descriptions (definitions) thereof. For that purpose, the descriptions are linked to data assumed to provide evidence and knowledge claims about them, their properties. Ontological meaning is tied to epistemic and methodological considerations and associated practices, experimental and theoretical. As a result, bio-ontologies can be said to categorize and classify both entities or phenomena and data (to interpret or categorize them), as a body of evidence for them.

Categorizations practices in such cases form networks of diverse kinds of asymmetric ordering relations (hierarchies) between described entities or properties: for instance, predication (classification), parthood (structural description) and causal regulation (causal explanation). While the information might seem static and cumulative, the practice of coordination of data sets and associated concepts is also dynamical in that it allows for the careful extension of the application of terms as well as for their revision.

The links between categorizations, or classificatory terms, provide then a dynamical integration of data sets. Through their interpretation, they constitute a form of higher-order knowledge claims about entities some consider a theory:[41] a set of testable theoretical hypotheses with meaning and validity established by empirical evidence and the reliability of the condition of its production. In that sense they may be considered both theory-laden and theory-making, but in the sense that classifying or categorizing may be a form of theorizing, constituting theory, playing a theoretical role (it's what theory and theorizing can be) and not just generating theory or theorizing.[42] Epistemic and methodological functions commonly associated with theorizing may be considered displayed by classification projects such as bio-ontologies: the roles of generalizing, unifying, explaining and research guidance.[43] Accordingly, classification projects may embody descriptive

---

[38]Leonelli [39].
[39]See Elgin [44], Cat [13, 40].
[40]Leonelli [39].
[41]Leonelli [41].
[42]Leonelli [39].
[43]Ibid.

and evaluative criteria applied to available data; they may also guide generalization and application of results and the pursuit of new empirical research, albeit in fallible, context-dependent ways.

Conditions of evidence and reliability of data are not merely theoretical and inferential. In data-driven empiricism, reliability of empirical evidence is not evidence of reliability. That is, different levels of evidence and reliability are in operation, also concerning different kinds of items, namely, general experimental techniques and apparatuses, and claims about particular phenomena and data sets; the last two kinds of claims are themselves related but not identical or equivalent. Yet the reasoning involved and assessing the evidential role of data (and mutatis mutandis the explanatory value of phenomena) is much like the one involved in assessing its reliability. Insofar as any form of reasoning is at play, it is often independent of the abstract theory that aims to explain phenomena and is empirical.[44] Reasoning doesn't operate alone, it is also embedded in the material, cognitive and social conditions of the practices involved in designing and constructing experiments and producing and interpreting data gleaned from them. Reliability of reasoning, making correct inferences about phenomena from data, depends on the reliability of data.

Of course, correctness is conditioned by two standards, one of conceptual precision and another of exactness. Reliability of measurement over a range of possible values or, more simply, detection of occurrence in contrast with its failure, is the outcome of exploring alternative processes of data production and making counterfactual assumptions such as that the inferred phenomenon (or associated categorization) would not have contributed to different data, and that the latter would have otherwise pointed to a different explanatory phenomenon. Knowledge of error characteristic of the detection or measurement procedure becomes relevant; and it is an empirical and probabilistic matter concerning the design, properties and behavior of the measurement apparatus and the standards of interpretation adopted, that is, the calibration standards for exactness and precision. Reliability is not established by simple calculation, or dictated by either a general theory of phenomena or a general theory of the apparatus. Reliability of data and reliability of an apparatus or technique are inseparable.[45]

Thinking about the reliability and interpretation of data is no purely logical or mathematical matter. Equally contextual is the preference for avoiding false positives over false negatives; this rests on another, namely, that the projects of avoiding the different kinds of misrepresentations stand in conflict. The relationships of evidence and reliability that apply to data might take the form of a simple rule for a task at hand, but they are interpretation strategies based on factual, complex and contextual forms of interaction.

What I have been describing thus far provides the grounds for data processing: the manipulation of reliable data into reliable and meaningful information.

---

[44]Bogen and Woodward [35], Woodward [36].
[45]Woodward [36].

Constraints from past empirical relations between partial data sets (indicators) and outcomes help interpret data more realistically in terms of evidential value and predictive reliability. Survey data provide a case in point, most famously voting data. Survey subjects become data production instruments of limited reliability. For instance, voters for a particular party may be historically less vocal than represented in election results. Processing the raw survey data for that particular party entails correcting accordingly the projected results. A similar constraint may be based on relations between indecision and abstention. In such cases, projection, a form of processing, involves contextual, empirically constrained re-categorization, or interpretation, of raw data.

Now, data are either visual or visualized, and from a functional standpoint of pictorial empiricism it's straightforward then that the considerations above apply to images as visual data, visualized data, and its higher-level, extended interpretations or categorizations. What I have called a content-extending rule is just the small and hardly stable tip of an iceberg of diverse practices, judgments, skills, habits, interests and other conditions and constraints.

An example is the regulative dimension tacitly present or expected from visualization as a practice. We may speak of the ethos of data visualization. Visualization, production and selection of data are inseparable. Pictorial visualization may be equally said in turn to involve inseparable activities of production and selection of its own. Again, then, evidence and interpretation are subject to regulative constraints, methodological standards, epistemic and ethical, rules of graphic integrity.

Admittedly, graphics can lie. They are not just approximately correct.[46] As Tufte has long warned, deception becomes a matter of scale and standard, when in that sense graphics can misrepresent the quantitative information contained with additional, often conflicting or unjustified, information conveyed only graphically. The extra information is an active form of cognitive intervention, based on interests and standards and leading to conflicts with epistemic virtues, or else by omission. Either way, such practices imply an added relevant dimension of complexity in the production, selection and interpretation of pictures. Both intrinsic and extrinsic forms of categorization play a role. It's not a mere matter of applying a rule expressing a formal inference or an empirical hypothesis.

Tufte had offered six principles of graphical integrity:[47]

(1) 'The representation of numbers, as physically measured on the surface of the graphic itself, should be directly proportional to the numerical quantities represented.'
(2) 'Clear, detailed, and thorough labeling should be used to defeat graphical distortion and ambiguity. Write out explanations of the data on the graphic itself. Label important events in the data.'
(3) 'Show data variation, not design variation.'

---

[46]Tufte [42], 76.
[47]Ibid.

(4) 'In time-series displays of money, deflated and standardized units of monetary measurements are nearly always better than nominal units.'
(5) 'The number of information-carrying (variable) dimensions depicted should not exceed the number of dimensions in the data.'
(6) 'Graphics must not quote data out of context.'

Insofar as data processing involves practices of categorization, it includes the possibility of fuzzy categorization and fuzzy visualization, although in a context of constrained cognitive, experimental, methodological practices.

References

1. Baigrie, B. S. (Ed.). (1996). *Picturing knowledge*. Chicago: University of Chicago Press.
2. Larkin, J. H., & Simon, H. A. (1987). Why a diagram is (sometimes) worth ten thousand words. *Cognitive Science, 11*, 65–99.
3. Daston, L., & Galison, P. (2007). *Objectivity*. Cambridge, MA: Zone Press.
4. Hentschel, K. (2001). *Mapping the spectrum*. Oxford: Oxford University Press.
5. Perini, L. (2005). The truth in pictures. *Philosophy of Science, 72*, 262–285.
6. Goodwin, W. (2009). Visual representation in science. *Philosophy of Science, 76*, 372–390.
7. Kulvicki, J. (2014). *Images*. New York: Routledge.
8. Frigg, R., & Hartmann, S. (2010). Scientific models. In E. N. Zalta (Ed.), *The Stanford encyclopedia of philosophy* (Winter 2015 Edition). http://plato.stanford.edu/archives/win2015/entries/scientificmodels/
9. Giere, R. (2001). *Science without laws*. Chicago: University of Chicago Press.
10. Bailer-Jones, D. (2009). *Scientific models in philosophy of science*. Pittsburg: University of Pittsburg Press.
11. Weisberg, M. (2013). *Simulation and similarity*. New York: Oxford University Press.
12. Lynch, M. (1985). Visibility. *Social Studies of Science, 15*(1), 37–66.
13. Cat, J. (2015). An informal meditation on empiricism and approximation in fuzzy logic and fuzzy set theory: Between subjectivity and normativity. In R. Seising, E. Trillas, & J. Kacprzyk (Eds.), *Fuzzy logic: Towards the future* (pp. 179–234). Berlin: Springer.
14. Hesse, M. B. (1966). *Models and analogies in science*. Notre Dame, IN: University of Notre Dame Press.
15. Gentner, D., Holyoak, K. J., & Kokinov, B. N. (Eds.). (2001). *The analogical mind. Perspectives from cognitive science*. Cambridge, MA: Bradford Books.
16. Hofstadter, D., & Sander, E. (2013). *Surfaces and essences: Analogy as the fuel and fire of thinking*. New York: Basic books.
17. Dunn, G., & Everitt, B. S. (1982). *An introduction to mathematical taxonomy*. Cambridge: Cambridge University Press.
18. Cat, J. (2001). On understanding: Maxwell on the methods of illustration and scientific metaphor. *Studies in History and Philosophy of Modern Physics, 33B*(3), 395–442.
19. Peters, J. F., & Pal, S. K. (2010). Cantor, fuzzy, near, and rough sets in image analysis. In J. F. Pal & S. K. Peters (Eds.), *Rough Fuzzy Image Analysis. Foundations and Methodologies* (pp. 1–15). Boca Raton, FL: CRC Press.
20. Dubois, D., & Prade, H. (1980). *Fuzzy sets and systems. Theory and applications*. New York: Academic Press.
21. Klir, G. J., & Yuan, B. (1995). *Fuzzy sets and fuzzy logic. Theory and applications*. Upper Saddle River, NJ: Prentice Hall.

22. Shin, S.-J. (2012). The forgotten individual: Diagrammatic reasoning in mathematics. *Synthese, 186*, 149–168.
23. Shah, P., & Miyake, A. (Eds.). (2005). *The Cambridge handbook of visuospatial thinking*. Cambridge: Cambridge University Press.
24. Kosslyn, S. M. (1994). *Image and brain*. Cambridge, MA: MIT Press.
25. Taylor, H. (2005) Mapping the understanding of understanding maps. In P. Shah & A. Miyake (Eds.), *The Cambridge handbook of visuospatial thinking* (pp. 295–333). Cambridge: Cambridge University Press.
26. Shah, P., Freedman, E. G., & Vekiri, I. (2005). The comprehension of quantitative information in graphical displays. In P. Shah & A. Miyake (Eds.), *The Cambridge handbook of visuospatial thinking* (pp. 426–476). Cambridge: Cambridge University Press.
27. Barwise, J., & Etchemendy, J. (1989). Information, infons and inference. In R. Cooper, K. Mukai, & J. Perry (Eds.), *Situation theory and its applications I, 1* (pp. 33–78). Stanford, CA: CSLI.
28. Barwise, J., & Hammer, E. (1994). Diagrams and the concept of logical system. In D. M. Gabbay (Ed.), *What is a logical system?*. New York: Oxford University Press.
29. Coliva, A. (2012). Human diagrammatic reasoning and seeing-as. *Synthese, 186*, 121–148.
30. Resnik, M. D. (1997). *Mathematics as a science of patterns*. Oxford: Clarendon Press.
31. Giaquinto, M. (2007). *Visual thinking in mathematics*. Oxford: Oxford University Press.
32. Cat, J. (2006). On fuzzy empiricism and fuzzy-set models of causality: What is all the fuzz about? *Philosophy of Science, 73*(1), 26–41.
33. Siegel, S. (2010). *The contents of visual experience*. New York: Oxford University Press.
34. Trillas, E., & Uturbey, L. A. (2011). Towards the dissolution of the Sorites paradox. *Applied Soft Computing, 11*(2), 1506–1510.
35. Bogen, J., & Woodward, J. (1988). Saving the phenomena. *Philosophical Review, 97*, 303–352.
36. Woodward, J. (2000). Proceedings of the 1998 biennial meetings of the philosophy of science association. Part II: Symposia papers. *Philosophy of science, 67*, Supplement, S163–S179.
37. Leonelli S. (2015). What counts as scientific data? A relational framework. *Philosophy of Science, 82*(5), 810–821.
38. Cat, J. (2013). *Maxwell, Sutton and the birth of color photography. A binocular study*. New York: Palgrave-Macmillan.
39. Leonelli, S. (2013). Classificatory theory in biology. *Biological Theory, 7*, 338–345.
40. Cat, J. (2016). The performative construction of natural kinds: Mathematical application as practice. In C. Kendig (Ed.), *Natural kinds and classification in scientific practice* (pp. 87–105). Abingdon: Routledge.
41. Leonelli, S. (2012). Classificatory theory in data-intensive sciences: The case of open biomedical ontologies. *International Studies in the Philosophy of Science, 26*(1), 47–65.
42. Tufte, E. (2001) *Visual Display of Quantitative Information*. Cheshire, CT: Graphics Press.
43. Tufte, E. (1991). *Envisioning Information*. Cheshire, CT: Graphics Press.
44. Elgin, C. Z. (1997). *Between the Absolute and the Arbitrary*. Ithaca: Cornell University Press.

Chapter 6
Visual Representation: From Perceptions to Pictures

I have pointed to the rather obvious fact that pictures, unlike linguistic symbols, have a specifically and distinctively visual character that is central to its uses. That is, pictures, like words, play many roles, as does categorization, not just representation, and for the performance of their functions pictures rely directly or indirectly on their perceptual properties. In this case, we can locate the conditions for vagueness of representation in the categorization of perceptual entities and properties.

The emphasis on perception suggests distinguishing between perceptions and pictures, and exploring them separately. We must simply recall that among perceptual entities are, of course, pictures. This strategy will help identify similarities and differences between perception and the perception of pictures and this in terms of at least four possible varieties of potential indeterminacy or vagueness: seeing fuzziness, seeing fuzzy, seeing a fuzzy picture and seeing a picture of a fuzzy system. It will also help identify the variety of kinds of contents of categories—and pictures—intrinsic and extrinsic.

Moreover, the distinction between perception and external pictures is not without entanglements involving the perceptual situations above. For instance, the intrinsic content of pictorial representations overlaps with the extrinsic content of perception, insofar as depiction involves actual or potential perception of the pictorial marks. In addition, in ordinary and scientific life technologically assisted vision extends the notion of perception to include the perception of external pictures. In such situations perception of visual displays of detected events or systems is often the only operative notion of observation, whether of microscopic, telescopic or otherwise inaccessible mesoscopic systems such as internal organs, geological structures, etc.

The subject displayed in a picture, and thereby constituting its extrinsic content, might not be unconditionally observable; in other words, its recognition requires perceptual assumptions. We can visually recognize a *variety of kinds of extrinsic*

© Springer International Publishing AG 2017
J. Cat, *Fuzzy Pictures as Philosophical Problem and Scientific Practice*,
Studies in Fuzziness and Soft Computing 348,
DOI 10.1007/978-3-319-47190-7_6

contents, e.g., referents or indexical targets, thematic descriptions, intended communications and inferred information. Their expression in categorization exhibits the active contextual role of viewers as cognitive agents engaged in a perspectival activity. It is a separate issue how the different properties associated with different categorizations are objective features of the systems described or at least items in the empirical world. This objectivity is compatible with another feature of categorization; properties as categorized or defined might be also relational, that is, exist within a context in which they can be recognized and be recognized accurately.

About perception, then, including perception of pictures: How rich can the content of perception be? How fuzzy? The answers to these questions are connected and help motivate further some of the ideas above. I turn first to Siegel's view of perceptual content in the absence of truth of beliefs or propositions. On her view, content is established by the scope of accuracy conditions and these extend to the representation of kinds. Perceptions are states that can be accurate: 'the notion of representation is tied to the idea that experiences have contents, where contents are a kind of condition under which experiences are accurate, similar in many ways to the truth-conditions of beliefs.'[1] In a deflationary sense of accuracy or error, talk of accuracy/inaccuracy implies the possibility of match/mismatch with some intentional content.[2]

In visual perceptions, images present properties as instantiated. The assumption of instantiation of categories is not only an explanation of perception and recognition in general. One may distinguish a non-conceptual, demonstrative images in perception, 'that F', from categorized images in truth-functional propositions of the form 'that's an F'.[3] But is the demonstrative character of percepts category-independent? That is, beyond linguistic labeling, is the demonstrative function independent of categorization (through the activation of recognitional capacities or mechanisms)?

Siegel emphasizes how we recognize properties besides any objects we attribute them to. On occasion, we claim to recognize only a property, even at a place or time, but this case is not a regular occurrence in ordinary perception.[4] Thus, a condition of accuracy of representation—that is, accuracy of images—is that there is something—an individual event, particular or, generally and loosely speaking, an object—instantiating the same properties. The possibility of content is the accuracy condition,[5] and, on the view I explore here, this is a condition on categorization.

---

[1]Siegel [1], 4. How this thought might apply to mathematical propositions or structures requires careful consideration of abstractions, formal truth by construction or inference.
[2]Ibid., 30–34.
[3]This is Ned Bloch's view.
[4]Ibid., 47. Here I'm glossing over many relevant and often controversial subtleties.
[5]Ibid., 45.

The individual/property duality captures a dual nature of content that is typically identifiable both in perception and depiction. This is no naïve realism about intentional objects: the perception of properties or kinds includes the typically relational, perspectival presentation of objects or events. These properties may even include what Siegel calls perspectival connectedness, the subject-dependent presentation of subject-independent objects,[6] e.g., relational properties involving relative motion or position, or action-readiness. This is the reason hallucinations can surprise us and illusions can fool us. The commitment to instantiated properties and kinds extends especially to causal categories and relations; they are particularly relevant, pre-theoretically, to the relation of perception to potential action and, theoretically, to the explanation and evaluation of perception and its reliability.[7]

From accuracy Siegel distinguishes what she calls weak veridicality, the property of experiences we deem veridical without being veridical of any object.[8] Weak veridicality applies, for instance, to hallucinations and illusions. We may consider this the phenomenological level of presentation. I suggest it is best understood as a case of intrinsic conceptual precision. In the case of intrinsic phenomenological properties that do not involve attributions to real things instantiating them, we may call this intrinsic content. On my reading, it is the extrinsic added level of representation that is the bearer of strong veridicality or accuracy. The intrinsic level of content or presentation supports the extrinsic level of representation.

Can experiences have both weak, or intrinsic, content and strong, or extrinsic, content, namely, one determined by its accuracy conditions? I think so, as a matter of ordinary cognition. We can recognize features of our perception at the expense of their accuracy; we also recognize them, both as potential signs of unreliability and of cognitive cues, on the basis of background information and beliefs. Both kinds of contents rest on an ability to categorize.

The scope of cases of weak content extends beyond ordinary success cases of functional visual perception and cases of visual malfunction: it reaches the overlapping domains of external images, technologically-enhanced perception and depiction. In those domains its role becomes more prevalent and it becomes an interpretive challenge to reduce representationalism, at least in the sense of the possibility of representation awareness, to direct realism, that is, to full transparency to the appropriately independent object of representation.

Now, the relation of representation in the form of conditions of accuracy includes the two cases of indeterminacy in the linguistic context where the proper understanding of vagueness has been originally debated. If approximation is the positive reading of the failure of accuracy, the failure of univocality may be of two kinds: single-target indeterminacy, such as vagueness, understood as a manifest semantic relation or objective property, and multiple-target indeterminacy, such as ambiguity. Representation in visual experiences may have accuracy conditions, but

---

[6]Ibid., 176.
[7]Ibid., 117.
[8]Ibid., 36.

they do not necessarily determine their target. It is not just an epistemic matter, it is a matter of lack of determination that causes our uncertainty; we are not sure of what we are seeing; our recognition ability is limited by multiple possible categorizations, but neither is determinate (unlike in cases of pluri-determination); identification is in each possible target system or representation, indeterminate.

Reference

1. Siegel, S. (2010). *The contents of visual experience*. New York: Oxford University Press.

Chapter 7
Vague Pictures: Scientific Epistemology, Aesthetics and Pragmatics of Fuzziness; From Fuzzy Perception to Fuzzy Pictures

A putative form of vagueness in perceptual presentation/representation is blur. From the standpoint of accuracy—in representationalism or intentionalism—blurred vision is like hallucinations, an instance of inaccuracy,[1] of misrepresentation. Things are not the way they are perceived or represented as being. Alternatively, it has been argued that along with error, blur has its own distinctive phenomenological presentation, in a broader sense, and recognition.[2] For authors such as Crane, Pace and Smith, blur challenges the transparency of visual representation and its content.[3] An alternative view to misrepresentation or underrepresentation is that the phenomenological property of blur is a form of overrepresentation.[4]

My view of blur's representational value—and content—differs in a way that is more pronounced in cases of technologically-assisted modes of perception and material-medium-based modes of depiction. In the generalized sense of categorization and objectivity that includes the ontic and the cognitive, and the semantic in between, we may think of meta-representation.

The overrepresentation interpretation of blurred perception and the argument for it are based on the assumption of accuracy conditions for representation. In particular, they assume our capacity to recognize specific features of the phenomenology of the blur, the tonal extension and variation and the extended geometry of a boundary. Still, whether a case of inaccuracy relative to sharp object or accuracy relative to fuzzy object, it follows that fuzziness is phenomenologically recognizable alongside any object whose putative perception is at stake. One may consider it, with Allen and others, a way of seeing, without distinctive content. But one may also explore the possibility of an added kind of content; this intrinsic content is its added categorization.

---

[1]Ibid., 49.
[2]Smith [1].
[3]Crane [2], Pace [3], Smith [1].
[4]Allen [4].

© Springer International Publishing AG 2017
J. Cat, *Fuzzy Pictures as Philosophical Problem and Scientific Practice*,
Studies in Fuzziness and Soft Computing 348,
DOI 10.1007/978-3-319-47190-7_7

Since this kind of phenomenological character, this way of seeing, rests on categorization—e.g., features of presentation—both misrepresentation and over-representation suggest a use for the distinction between what I call intrinsic and extrinsic contents. In a narrower sense of representation, overrepresentation would seem to deny the distinction, when in fact it only shows the important fact that both contents are related. The relation is neither fixed nor general. This is why understanding pictorial indeterminacy matters. The role of the distinction and the relation between both kinds of contents becomes more pronounced and relevant in the case of pictures.

It is not surprising that Allen, for instance, emphasizes the distinction between blurred images and pictures. In the case of intrinsic content, limited to visual phenomenology, I assume an extended notion of content that concerns the conditions of conceptual precision of visual categorization, and yet another distinction, between accuracy-content—with Siegel and Allen—and precision-content. Only the latter is intrinsic.

Each kind of content allows for different forms of indeterminacy and vagueness; they are related by varieties of IC–EC links, e.g., opaque links, explicit rules, or hybrid cases. Visual recognition is, on the surface at least, a mode of interpretation in the form of articulating content through categorization; and it may proceed by construction, through the exercise of different skills: inferentially, by explicit rules or transparent algorithms, which we learn to apply in different contexts, sometimes very much linked to a particular type of situation, format, medium or technique; or more unconsciously through cognitive activities whose processes we cannot fully understand or control, including visual cue-type associations; or hybrids such as imaginative activities of interpretation providing complementary representation.[5]

As an account of blur, overrepresentation is inadequate. In defense of representationalism, it seeks primarily to avoid and analyze away indeterminacy or vagueness, instead of adequately incorporating it on its own terms. To the extent that Allen's account incorporates it, it does so by identifying vagueness with an instance of truth-glut, a bundle of simultaneous representations with inconsistent information or content. This view is a controversial alternative to the notions of indeterminacy and partial-truth view of linguistic vagueness I am examining here and have motivated at the start; especially, when the criterion of vagueness is based on objective conditions of closeness and partial-membership or categorization. In particular it fails to capture the relevant kind of truth and representation; it also fails to capture distinctive features of blur and fuzziness. They might count as cases of overrepresentation, but only to the extent that they include over-categorization and vague categorization, whether intrinsic or extrinsic. The only truth-bearing

---

[5]Zadeh's distinction between opaque and transparent algorithms is, as is every distinction, too simple. This diversity is relevant to different types of situations and features, for instance, in cognitive approaches to the viewing and understanding of film images. Interestingly, the recognition capacity that is exercised in film viewing along with any acquired competence in the conventions and choices of authors, genres or styles is itself based on combinations of perceptual disabilities; see Thomson-Jones [9].

representation is an associated linguistic one, which arguably includes an element of partial truth.

In blurred perception, a figure's boundary often extends into a narrow continuous surface. This presents two challenges. The overrepresentation account assumes that this extended boundary region is a composite of determinate line-type representations or categorizations. But whereas the representation of the region is visual, the alleged line components often are not; and visual, not mathematical, representations are the only ones of relevance. In addition, the boundary region marking or enveloping the figure has itself a boundary, which often lacks sharpness or definition; it is diffused and resists being categorized in terms of a sharp line.

Over-categorization of the boundary of a figure, instantiated or not by an object, cannot be fully determinate if it aims to describe the phenomenological state. The same applies to the categorization of any other kinds of fuzzy experiences we call diffused or hazy. If overrepresentation assumes determinacy of extrinsic content, it must also assume inaccuracy. Beyond that, it must assume vagueness.

If indeterminacy is only of the extrinsic kind, it is complex. Blur, on my interpretation, includes both kinds of indeterminacy mentioned above. The boundary region is a region of indistinguishability (the kind formally modeled by rough sets), and it leaves indeterminate possible sharp line boundaries and higher-order ulterior categorizations associated with topological properties. The diffused surrounding area is the source of vagueness, a hazy edge with a tonal variation that prevents the perception of a line based on a sharp tonal contrast. It is the kind of situation formally modeled analytically with gradients and set-theoretically with fuzzy sets and indistinguishability-based near sets. The borderline is a borderline case of geometrical—topological—recognition through categorization. Vagueness may in turn be an additional basis for pluri-indeterminacy. This is cognitively the case whether vagueness is an extrinsic property or an intrinsic categorization independent from the accuracy conditions involving interactions with environmental systems or events. Still, to the extent that perception of haze is amenable to categorization or recognition in terms of the experience, and not of instantiation by a system, blur may have an intrinsic content and status. While the possibility I am formulating is conceptual, it establishes an empirical possibility; it is then a contingent empirical matter whether the type of situation is empirically actualized or instantiated and what additional features may become relevant; this is a place for further discussion.

Whether as a tenet of representationalism or a factual case, the contextual distinction between intrinsic and extrinsic contents may collapse without eliminating the various relations between visual vagueness and indeterminacy of perception. Indeed, this is the basis for the kinds of instances of the distinction one can find in cases of depiction by (perceived) pictures. Precise representation may have a univocal and accurate content; it may still be indeterminate contents in other ways. Alternatively, vagueness of categorization, intrinsic or extrinsic, may support both determinate and indeterminate contents. But neither case is one of overrepresentation.

The analysis I have presented is abstract in the sense that it is largely independent from a variety of more specific perceptual situations and phenomena, and their significance, and presumably compatible with most of them. How the relevant phenomena of perception and pictorial representation are best modeled is, from the point of view of their cognitive psychology, complex and controversial. But the differences pose welcome opportunities for the application of fuzzy theory and technologies of image processing. Should the application of fuzzy theory track the additional level of specificity and detail? This is generally a matter of empirical concerns; but it might be more directly relevant to the project of technological application if its aim includes reproducing actual mechanism of perception and recognition realistically—that is, according to some empirically adequate model— or else heuristically, insofar the simulation helps achieve a technical goal.

I will not pursue the issue here. Nevertheless, in the next chapter I widen the focus to consider the relation between the cognitive and aesthetic significance of perception and pictorial representation in painting and photography; one kind of phenomenon and significance draws attention to the other, extending the scope of significant fuzziness. The root problem is the challenge to the old cognitive standard of the innocent eye and its pictorial counterpart in the arrested image. The visual field provides no sharp view from nowhere; the visual representation it constitutes is instead radically situated and embodied, conditioned by multiple anatomical, physiological and environmental circumstances.

In the opposing standard the visual field is the product of a vision process that is contextual, centered, dynamic and constructed. The characteristics are inseparable, one enabling or constraining the others; and each can support specific formal and empirical models. The contextual dimension includes environmental cues and conditions such as relative motion, invariance, distance and atmospheric and lighting conditions. This embedded condition has been the object of influential twentieth-century analyses as different as Gestalt psychology, von J. von Uexküll's model of enveloping *Umwelt*, Merleau-Ponty's holistic phenomenology and J. J. Gibson's subsequent ecological approach.[6]

The centeredness of the field of view has been a recurrent feature in eye-centered geometric models of vision since early Greek theories such as Euclid's and subsequent development in the form of systems of perspective such as Alberti's for the representing perceived size as a function of distance. The learned experience of systematic distortion becomes a cue that informs us about distances. One key added feature was binocular three-dimensional vision through the coordinated control of moving eyes.

Another, more relevant to this chapter, is the selective focus of attention, whose features were compiled in Helmholtz's writings in the late nineteenth century. The focus presents at least three related cognitive aspects: triggering environmental cues, higher-order conceptual interests and background information, and, at the

---

[6]Gibson [5].

phenomenological level, a localized increase in clarity—that is, resolution, discrimination or sharpness.

The last two are the most relevant to the identification of fuzziness, especially to the extent that together they involve different levels of categorization, more phenomenological and more abstract—more intrinsic and more extrinsic. Attention is a constrained cognitive state and activity, driven by vigilant and probing attitudes. The first of the three aspects is not irrelevant; for instance, the sensitivity to motion and its role in guiding our vigilant attention illustrate how which peripheral vision is not just background texture with low resolution, but a partly unattended, partly informative visual field. A relation to the other two aspects is evident first in the role of conditions of attention that inform the field, with weak awareness or even what Michael Polanyi called tacit awareness; and second in the increase in clarity, awareness and informativeness that follow the shift in center of probing attention.

The last relation connects centeredness to the dynamic character of visual perception. At least three dimensions of this dynamic character are worth distinguishing. One is the roaming eye's higher-order probing as a practice of selective attention guided by cognitive expectations, interests and concerns.[7] Another is the random wandering micro-motion of the eyes and the head, a sort of background motion, unconscious and more frequent; this dynamic level contributes to the benefits of the first—indeed it may be considered a degree-zero version—and both might well have similar evolutionary explanations.

The third one is no less interesting and it also relates to the place of perceptual fuzziness in internal and external representations, in this case especially the lack thereof. Despite the motion of the eye's sharp centered focus of attention, internal representations appear as a uniformly sharp static image, either retrieved in memory or in "immediate" perception. The myth of the arrested eye and the corresponding arrested image has suggested the view that perception is not a basis for thinking but it is based on unconscious kinds of thinking. The processing of the continuous visual input or a sequence of time lapses yields a static composite that becomes a phenomenological object of awareness.

We can list several situations in which experience of the environment—the perceptual interaction with the environment in which we are embedded—has frequently been reported to feature some form of visual blur or fuzziness:

- Focal distance
- Physical distance
- Relative motion
- Environmental diffusion
- Selective attention

Fuzziness becomes cognitively significant as a cue about the properties of the environment or systems in it such as relative distance (Fig. 7.1) and motion (Fig. 7.2). It is also significant as part of the exercise of recognitional capacities in

---

[7]For a useful survey of more specific issues and situations see Gong et al. [6].

Fig. 7.1 Photograph of a country road with change in size and sharpness as distance cues. ©
George Clerk

limiting the focus of attention. In the case of perspective it is the distortion of
decreasing size that provides metric information; in this case, it is the decrease in
sharpness.

Environmental diffusion is an environmental effect that explains the perception
of fuzziness relative to the possibility of sharper experiences associated with theoretical or empirical information about a certain system. Informationally, we can
think of visual noise. For instance, atmospheric conditions may form a halo around
the bright image of the moon with diffusion that blurs the edges that, by experience
and astronomical information we assume to look sharper in more ideal conditions

Fig. 7.2 Photograph of people in motion with blur as motion cue. © George Clerk

(more ideal conditions are those in which interfering conditions that can explain physically the distortion are absent). Distance effects are not entirely independent from others; for instance, they are compounded by diffusion effects—this is noticeable in the resulting fuzziness of digital photographic pictures (Fig. 7.3).

For the purpose of discussion, it is crucial to keep in mind that, as these effects suggest, experienced sharpness and, by implication, fuzziness are always relative. Each set of conditions determines a standard. At the very least, the relevance of each listed effect is obviously relative to eyesight conditions, that is, optical functionality. Focal distance is the telling situation in which blur is associated with bringing the image of an illuminated object into optical focus; this experience sets the standard of sharpness. Again, also this effect can compound others. At the same time, focal distance provides the mechanism for selective attention (Fig. 7.3). In turn, at the perceptual level, selective attention might be prompted by other environmental cues such as motion.

The case of focal distance has an important cognitive extension for a spectrum of intrinsic and extrinsic features that we may a call critical distance effect. At some critical distance the environmental conditions trigger the recognition of certain visual feature such as the outcome of a color composition or some pattern of intrinsic features that we associate with familiar extrinsic content such the appearance of some kind of thing or symbol.

How the process of visual representation as a cognitive practice works is a problem tackled by a vast number of formal and technological research projects that in the last couple of decades have mainly developed Marr's computational

Fig. 7.3 Photograph of a cow's head in which focal distance serves selective attention, with maximum central sharpness and peripheral blur (in contrast with the central blur in the distance effect). © George Clerk

approach.[8] A desirable, but typically neglected, measure of empirical adequacy of the formal models is how they treat fuzziness. And the task is twofold, since it requires accommodating both the appearance of fuzziness in visual inputs and its disappearance from the "final" representation. The two are inseparable since fuzziness is most noticeable in more static experiences, whereas the shifting of the sharper center of attention seems to play the role of spreading sharpness in the process of the recognized visual experience. This is a task for fuzzy computational modeling; one strategy is to subsume the case of perception under more general models of learning that often target higher-order, more complex cognitive tasks; and in technological models of simulation and processing, under general fuzzy models of machine learning and pattern recognition, with non-Boolean algorithms incorporating different membership functions and operations that help formulate rules and conditionals (besides their place in rules, conditionals contribute also to the generation, regulation and assessment of approximate reasoning).

An alternative strategy is the application of fuzzy versions of formal treatments in models of dynamic vision models. Each treatment provides models with empirical and technological strengths and limitations. Like treatments of blur, they typically come in two related formats, analytic and statistical or probabilistic. An

---

[8]Marr [7]. Some of the research also adopts assumptions from Gibson's ecological approach. The notion of perceptual learning as a model of vision and cognition in general was effectively introduced by Helmholtz in the late nineteenth century; it was part of a project of naturalizing some of the tenets in Kant's epistemology.

example of analytic treatments is the geometric approach based on the so-called Gabor wavelets, special Gaussian treatments of spatial frequencies; the Gaussian distribution, while providing a solutions to wave equations also provides a statistical basis. The other statistical or probabilistic approaches are Principal Component Analysis (PCA) (and variants), Linear Discriminant Analysis (LDA), Gaussian mixture estimations (especially, expectation maximizations), Kalman filters, Bayesian belief networks and hidden Markov models.[9]

One strategy for their fuzzification is foundational, by developing the mathematical formalism from fuzzy versions of their respective set-theoretic basis. Efforts in this direction have focused on probabilistic and statistical techniques at the expense of analytic treatments. The new formalism requires also modifications in the inferences and other constraints we find in analysis and probability theory and statistics insofar as they may be based on incompatible classical assumptions. This strategy is not free from interpretive or conceptual challenges, since as a measure of uncertainty probability theory is based on Boolean structures of the random distribution of events. The probabilistic or statistical interpretation and reasoning it motivates disappear in the generalized formalism; and they are recoverable in some cases only in the classical limits for fuzzy and rough notions of information and entropy, also introduced as measures of uncertainty (see above). A more direct set-theoretic reformulation, as in the case of rough and near sets, may be introduced in the formalism of fractal set theory. All these suggestions and remarks require detailed technical analysis, which I cannot offer here.

Finally, it is easy to see how the third, dynamic, characteristic of visual perception leads us to the fourth, its constructed character. Recognition as a task of categorization has been considered from a number of viewpoints with a number of different models, often related, such as Gestalt laws of visual field organization, Popper's Darwinian cognitive psychology of trial and error in the form of falsifiable conjectures, information theory and prototype theories. The latter two rely mainly on reasoning and computation, including importantly analogical reasoning. The former are based mainly on a set of simple schemas and projection mechanisms. Where they differ is in the source and kinds of schema at work, expressing innate dispositions, neurological structures, or conceptual structures developed further by constraints from the environment, and, in more complex and abstract kinds of cognition, by pragmatic and conventional considerations.

References

1. Smith, A. D. (2008). Translucent experiences. *Philosophical Studies, 140*(2), 197–212.
2. Crane, T. (2006). Is there a perceptual relation? In T. Gendler & J. Hawthorne (Eds.), *Perceptual experience*. Oxford: Oxford University Press.

---

[9]See their application for instance in Gong et al. [6], Bishop [8].

3. Pace, M. (2007). Blurred vision and the transparency of experience. *Pacific Philosophical Quarterly, 88*(3), 283–309.
4. Allen, K. (2013). Blur. *Philosophical Studies, 162*, 257–273.
5. Gibson, J. J. (1979). *The ecological approach to visual perception.* Boston: Houghton Mifflin.
6. Gong, S., McKenna, S. J., & Psarrou, A. (2000). *Dynamic vision.* London: Imperial College Press.
7. Marr, D. (1982). *Vision.* San Francisco: Freeman.
8. Bishop, C. M. (2006). *Pattern recognition and machine learning.* New York: Springer.
9. Thomson-Jones, K. (2010). *Aesthetics & Film.* New York: Continuum.

Chapter 8
Blur as Vagueness: Seeing Images Vaguely and Seeing Vague Images; Perception and Representation

What is the point of examining the role of vagueness in perception? Pictures are visual items whose use depends on their perceptual properties. Their perceptual character can be established either through their generation, partly through eyesight, or their ability to be seen. Pictures can represent, among their possible uses, partly by virtue of their visibility. As a result they possess a twofold status, they can be seen and the viewer can also see in them or, imaginatively, through them—he can recognize actual objects of experience or conventionally encoded information, and be aware and even come to believe he does. What is the value of the distinguishing between perception and depiction, image and picture? Since one might consider the processing of mental images to constitute the ground zero of visual cognition, it is a relevant domain to examine in order to explore whether vagueness is possible in any form and role. With the linguistic standard of vague predicates as a starting point, this possibility becomes inseparable from the possibility of different kinds of contents linked to different categorizations. The case of blurred images is instructive because it challenges the emphasis on representation based on assumptions of precision and a single kind of content.

To insulate blurred vision from representation is the point of the two distinctions Allen wants to draw between intentional mode and intentional content of seeing and between blurred images and blurred pictures. Not unlike Siegel, he seeks to defend a representationalist, or pure intentionalist, view, in terms of the distinctive transparent content of images describing properties of objects in the environment, or at least ascribing the properties to them precisely—and more or less accurately or reliably, Siegel would add. Where Siegel relies on inaccuracy of such precise representations, Allen relies on their proliferation. A commitment to vagueness rejects their shared common assumption.

For his purpose Allen postulates that understanding blur in perceptual representation requires meeting two conditions, or challenges, involving pictures: (1) to explain an alleged difference, namely, to explain blur in images without reducing it

© Springer International Publishing AG 2017

J. Cat, *Fuzzy Pictures as Philosophical Problem and Scientific Practice*,
Studies in Fuzziness and Soft Computing 348,
DOI 10.1007/978-3-319-47190-7_8

to blur in pictures—as in the private screen model of the visual field—on the assumptions that only the picture case involves the existence of a representational medium and that blur can be explained by appeal to distributions of shapes and colors in it; and (2) to explain an alleged similarity, namely, the alleged phenomenal equivalence or indistinguishability between both kinds—between blurred images of determinate representations with object boundaries and clear images of fuzzy representations with object boundaries (Figs. 8.1 and 8.2; note that both figures are sharp pictures of the reproduced pictures, one fuzzy and the other sharp; the sharp image of a fuzzy picture simulates by analogy a fuzzy perception of a sharp object). Neither condition appeals to properties of the objects (realism), properties of the visual field (subjectivism) or to modes or degrees of seeing (impure intentionalism) to provide the best explanation of blur.

Both conditions, of distinction and of similarity (along with perceptual alertness in earlier analyses of realism in pictures and the possibility of inaccuracy), suggest an additional kind of representation—instead of additional representations. The image/picture dichotomy is neutral as to whether we should take the phenomenal kind as transparent (sharing relevant features with the represented object) or not, whether one wants to reject an opaque non-representational reading or a subjective dimension. It is also neutral as to what factors play a role in fixing representation

Fig. 8.1 Sharp image (on this page) of a fuzzy picture (photograph) of a sharp system, the moon. Moon over Goldmühl, courtesy of Kilian Heck

Fig. 8.2 Sharp image (on this page) of a sharp picture (photograph) of a fuzzy system, fog bank over a forest canopy. © George Clerk

and categorization. This is a matter of context and of object. If we follow Allen in adopting the distinction between blurred images and pictures, the fuzzy content of blurred images may still be understood in terms of a dualistic categorization: the phenomenal and the background-determined kind (the background includes at least a context of information and conventions, a framework of beliefs and a set of skills). This provides the unifying framework that helps makes sense of the presence of vagueness in pictures and its formal modeling. In particular, the phenomenal content or representation I associate with mere categorization may be playing the role of fuzzy intrinsic content of fuzzy pictures (see below).

Moreover, the emphasis on the distinction in condition (1) faces a challenge I have already mentioned, whenever perception is understood broadly, when its role in much of science and more general ordinary modern life is mediated by a material medium or technology. In that context, seeing is part of seeing a picture and seeing in a picture. From the standpoint of sources of information and the methodological use of data, we observe phenomena and measurement data on a variety of displays, e.g., dials, scales, digital counters, computer screens and TV monitors. The lesson from the case of pictures on vagueness and indeterminacy applies, more broadly, to representation in extended, technologically-assisted perception, where the screen and the visual presentation are not private but material and public.

In this framework of material conditions, we can understand better the contextual character of vagueness and indeterminacy. Background conditions that include mechanisms and beliefs inform our recognition of perception, or its contents. Whether the categorization of experience is intrinsically or extrinsically vague may be partly grounded in two entangled kinds of judgments: of accuracy and of reliability.

While the conditions of accuracy contribute semantic conditions, or the content, the conditions of reliability of the medium or mechanism contribute cognitive conditions. Whether we see a boundary line or we see it blurrily depends partly on theoretical beliefs and empirical habits concerning the environment, including geometrical and tonal concepts. Whether the shapes and colors we see correspond to properties of existing structures depends also on assumptions about the structures' properties.

Also play a role assumptions about the functioning of the part of the world that is the cognitive medium or apparatus of vision and visualization, whether anatomy, physiology or technology. This is the site of subjective-objective polarity. This second kind of background assumptions inform the reliability judgments that help determine categorization: we may find an experience intrinsically vague or indeterminate relative to expectations about specific determinate properties of the environment, e.g., the structure that should be perceived relative to us in a specific shape or figure, and the causal mechanism of its representation, or distortion, e.g., failures to meet standard optical, anatomical or physiological conditions. These may include chemically induced hallucinations, optically induced distortions or anatomically induced disabilities. One important type of case is the phenomenon of dependence on scales and levels of resolution, e.g., we are expected to identify the content sharply or in any other specific way from a certain distance, with a certain magnifying power etc. The situation extends to technological conditions.

The same kind of contextuality, as the discussion of perceptual and material realisms suggests, extends to the cognitive value of vagueness based on assumptions that allow us to recognize less self-consciously in the intrinsically vague properties, or infer from them, some determinate extrinsic content such as a depth cue or else a fuzzy textured surface (and in fact also assume the determinate content as the standard that leads to acknowledging some intrinsic, phenomenal indeterminacy). They are the same kinds of assumptions that help eliminate extrinsic indeterminacy in the indexical interpretation of a sign or symptom in relation to its explanatory cause—e.g., an underlying condition or disease—or a predictable causal relative—e.g., an upcoming economic or meteorological situation.

I find equally questionable, then, Allen's assumption that perception is both uniquely and precisely categorized, with the result that representation is precise and its content is invariably determinate, and that its extrinsic properties are presented accurately or inaccurately in precise representation or else in unreliable and inaccurate overrepresentation. The determinacy status of the intrinsic content doesn't generally and univocally fix the status of the extrinsic content.

As different examples suggest, the indeterminacy of extrinsic content extends to the possibility of vagueness. It might become challenging to identify what we are seeing and suggest a number of categorizations. Even if, despite indications to the contrary, we insist in denying intrinsic content to perception, extrinsic representation can bear both kinds of categorization and, therefore, of representation, and present vague and indeterminate in different combinations. If we accept both kinds, any state of determinacy at either level, intrinsic or extrinsic, may play a role in establishing the state of the other. Again, the specific character of their relation is not fixed, except in specific and varying ways according to varying sets of conditions or contexts.

Part of the occurrence of vagueness is the plurality—without sharp overrepresentation—of different sources of content and content indeterminacy. From the standpoint of categorization, such diversity may characterize visual experience as much as pictorial representation. Some of the sources of indeterminacy come in the form of (contextual precise enough) dichotomies between levels of content description, e.g., intrinsic/extrinsic, object/aspect, kind/particular, generic/specific, bare-bones/fleshed-out and whole/part. In turn, instances of each kind of categorization may apply vaguely.

Intrinsic description represents the more phenomenal properties of experience representable in geometric, topological or chromatic terms, e.g., seeing particular shapes, boundaries, patches of colors or degrees of brightness. Extrinsic description represents independent systems or events and their properties, e.g., seeing a black crow fly above one's head or a strengthening hurricane storm moving on a weather map. The intrinsic/extrinsic distinction has been formulated in terms of two related and equally contextual and imprecise dichotomies, syntactic/semantic properties and bare-bones/fleshed-out content.[1]

Object descriptions may present entities or events, e.g., a crow and a hurricane; aspect descriptions present features of an object, e.g., an aggressive crow and a strengthening hurricane. Then the object/aspect distinction can be associated with the object/subject dichotomy in the content of depiction, e.g., a hurricane is causally referred to (indexed), but is described as fast approaching a crowded city and a crow is referred to but presented as a virus carrier in a deadly epidemic.[2] Kind descriptions present a class of objects, often linked to a property, e.g., a crow or bird and a storm or hurricane; particular descriptions present an instance of a kind often denoted by singular noun, e.g., my pet crow Jimmy or hurricane Sandy.

Generic/specific descriptions present schematic, abstracted or simplified selections of features, e.g., a stick figure, a cartoon, a sketch of a landscape lines or patches of different color and brightness; they tend to carry more content than just the intrinsic visual properties; specific descriptions present more determinate, richer, detailed versions with specifics, what we would taken as more informative, more realistic, e.g., the image of a recognizable human body with anatomical features or a detailed portrait of a particular location.[3] The more properties

---

[1]Kulvicki [1]; the distinction is originally due to John Haugeland.
[2]Lopes [2].
[3]Hopkins [3].

represented, the more determinate the explicit content, thereby making the picture potentially more or less accurate, rather than indeterminate.

Regarding the part/whole distinction, and the many relations that express it, it is a familiar experience that descriptions of a whole (not a trivial judgment in any context) may identify something independently of the recognition of specific parts, e.g., a face; part descriptions identify composing elements, e.g., the shape and color of hair, eyes, lips, etc. To see a whole precisely does not require seeing its parts with the same degree of precision, intrinsic or extrinsic. This is a holistic feature of images; in analyses of perceptual realism it is often illustrated by reference to the focus of attention (see also Chap. 11, below).[4] We can identify someone's face precisely without identifying its parts; and vice versa, at the higher level of resolution, closer distance, larger magnifying power, etc., we see clearly details of the face without recognizing the face. Nor do we recognize jut any parts, for instance, we don't recognize parts we have not individuated uniquely before by association with the whole; we don't recognize a face on a digital screen by recognizing any or all individual pixels. Vagueness in the visual experience of parts is compatible with, if not a condition of, clear experiences of a whole, whether at intrinsic and more abstract or complex extrinsic representations. The same considerations apply to pictures. The whole/part relation involving vagueness is compatible with the Fodor-Sober condition that parts of pictures represent parts of the target object, even when parts aren't spatial, as in mental images.[5]

With an emphasis on the role of categorization, the kinds of realism I have mentioned share a semantic difference from the phenomenon of linguistic vagueness. Despite commentators' references to truth and truthfulness I discuss in Chap. 11 the operative conditions of veridicality, accuracy and representation are neither strictly nor recognizably species of truth relations. Accuracy and representation conditions for intrinsic and extrinsic contents are based on considerations of precision of categorization and, according to common linguistic uses, instantiation. The applicability of standards of accuracy implies a standard of precision; conversely, no inaccuracy without its precise measure.

If categorization provides a common framework for identifying vagueness in language and images, we can also point out differences.

(1) *Categorization in match without truth.* Smith's closeness condition for vagueness becomes a matter of predication rather than truth. This result modifies how we should understand the role of formal models in terms of the application of fuzzy and rough sets (and the tolerance conditions in the more general near sets). But their application remains possible wherever we identify a primary role for categorization. This is the reason we use captions to fix the meaning of an image or claim it illustrates a particular message communicated

---

[4]Schier [4].
[5]Sober [5].

by the text. The truth of a picture is strictly the derivative truth of a proposition about it, especially about its relation to its external content. And, as I have noted in earlier chapters, it is no trivial form of transcription. A similar claim has been argued about scientific models whether as visual or non-visual representations.

(2) *Categorization of representations as well as of their objects.* Another difference from the case linguistic representation, pictorial categorization applies to representation as well as to its extrinsic contents. In the linguistic case, one might argue, categorization applies only to the referent or subject of predication as indicated, denoted, by the predicate. Alternatively, one might argue that reading involves a similar task of categorization of intrinsic features of symbols, but that the latter stand in a different kind of IC-EC relation to their content than pictures or perceptions do (again, except perception of words).

Separate categorizations establish separate kinds of contents, intrinsic and extrinsic, especially external: again, we can describe tonal features of color variations and geometrical properties of lines and surfaces; and through them we can identify additional information or meaning. For instance, an intrinsic red light might be identified extrinsically as a state of emergency and a green surface with an irregular boundary may be identified as a region on a map, and so on for additional constructions of meaning or content associated with the preceding one by some process or procedure—and the same will apply whenever we admit only of extrinsic content, as in perception (above).

The role of any resemblance or matching condition will become a matter of establishing co-categorization or co-instantiation. Admittedly, relevant categories might be derivative from perceptual ones and the extrinsic content might be identified relative to the context of available experience, information and intended use. Thus, similarity conditions may admit of different precise formulations and specific determinations, including set-theoretic ones; what counts as similarity and what is the particular similarity or its degree are contextual matters linked to conventions, capacities and purposes.

Any truth conditions will be derivative, based on associations between statements and pictorial representation they refer to. In such representation (or presentation) statements, predication expresses categorization. Yet, I insist, this role of categorization is hardly exclusive or universal. The generality of this point becomes more evident in my discussion of depiction. Different kinds of images and pictorial practices and situations involve additional elements, even in the case of the different practices of categorization, including the activation of opaque mechanisms of recognition.

This has consequences for our understanding of visual representation and reasoning, the role of various kinds of indeterminacy in them and in the application and understanding of fuzzy theory. So, we need to ask how categorization in the generalized sense presented above can make vagueness possible in pictorial representation. My discussion so far has provided and motivated elements for an answer. Understanding vagueness in visual experience contributes the framework for understanding the case of pictures; in turn the case of pictures extends the conception of perception and its understanding.

References

1. Kulvicki, J. (2014). *Images*. New York: Routledge.
2. Lopes, D. (1996). *Understanding pictures*. Oxford: Oxford University Press.
3. Hopkins, Ch. (1998). *Picture, image, and experience*. Cambridge: Cambridge University Press.
4. Schier, F. (1986). *Deeper into pictures*. Cambridge: Cambridge University Press.
5. Sober, E. (1976). Mental representations. *Synthese, 33*, 101–148.

Chapter 9
Vague Pictures as Pictures

In this chapter I use the notions introduced above to lay out conditions of vagueness in pictures, the diversity of roles and scenarios and differences from vague linguistic predicates and representation.

As a matter of accuracy of contents, we can distinguish vagueness from *approximation*: the latter includes an asymmetric relation to a fixed precise value and a measure of incomplete or partial characterization; although we might not reach the exact "target" representation, as in the exact value of π, or knowledge of the full set of properties or a any standard of completeness; in the last section, below, I examine approximations in more detail and argue that fuzziness reduces vagueness to approximation.

Given the constraints on our conditions of perception, representations of three-dimensional objects seen from a particular perspective and distance, etc. onto a two-dimensional surface (including the retina), all pictures are in principle partial representations. In this sense, they present a structural representation deficit. Incompleteness or inaccuracy may be considered forms of conceptual approximation. They may also exhibit a degree of representation excess insofar as some of their properties in combination or separate from the ones shared with the object of representation or relevant to its representation may be said to represent other objects, real or abstract.

Neither perception nor representation is fundamental. Representation is only one of the uses and roles of pictures as visible surfaces in a medium. Categorization itself is only one of the many elements contributing to representation (among other functions, some non-representational). I have focused on their relation, as I have indicated above, as a heuristic that helps explore the value of the standard set by the linguistic case and test its limits.

I have been examining conditions of vagueness of pictorial representation as a property of a specific relation of depiction and its content. But representation is established and understood in a number of ways, alternatively or in combination, in different contexts, with a varying range of applications. Vagueness is a feature of a role established by multiple criteria such as use, recognition, resemblance

© Springer International Publishing AG 2017
J. Cat, *Fuzzy Pictures as Philosophical Problem and Scientific Practice*,
Studies in Fuzziness and Soft Computing 348,
DOI 10.1007/978-3-319-47190-7_9

(perceived or including structural mapping), causal indexicality, information, intention, pretense, tools or conventions, cognitive capacities, competence, and so on.[1] Each criterion is subjected to considerations of different representational practices. Defenders of each criterion typically assume its exclusive value in establishing representation. I take a more pluralistic standpoint about how representation may be understood and performed in different contexts. Not only different criteria may be relevant to establishing and understanding representation; in the same context they may do so also jointly. There are many specific ways fuzzy categorization and representation can be established (implemented or warranted).

Placing an emphasis on perception—especially seeing, but not exclusively or alone—narrows the range of ways categorization enables depiction. Also, as I discuss next, it allows for a distinction between two elements, a pictorial object and its associated object, and show how they are both subjects of categorization. They are endowed with corresponding kinds of contents in ways uncharacteristic of linguistic representation, and they are nevertheless able to manifest corresponding kinds of vagueness, in different possible combinations.

Since it's not my purpose to defend a general account of depiction, I will not worry about potential counterexamples of the sort like pictures without designers, producers or users, invisible representations, sensory deprived makers or users, or hypothetically unseen objects that might bear some resemblance or causal relation to others in a purely theoretical or metaphysical sense… in the desert or while we sleep.

From this perspective, to use a picture is to use a visible surface in a material medium of support and display by engaging four kinds of perceptual experience: (1) seeing a picture and its properties, (2) seeing something else in it (and in the process, the feeling of seeing through it), (3) seeing the target system and (4) seeing other images of it.

Appealing to the concept of seeing-in provides a criterion of pictorial representation in terms of an experience that triggers a conceptual response: a judgment of identification in terms of a particular or general concept, individual or property. Seeing-in offers a criterion adequate to the perceptual nature of pictures. The idea is typically associated with Wollheim's twofoldness account of depiction; the view takes to be distinctive and irreducible two perceptual features of pictures: that we perceive them as objects with intrinsic properties, and that we have Gestalt-like experiences of aspect-seeing, or seeing-as, seeing the picture as the putative depicted object, the extrinsic properties.[2] Thinking of pictorial depiction in terms of the experience of seeing-in is a way to take seriously the role of perception in its two forms.

As I have noted above, what I call the intrinsic content is simply part of the external content of perception. What about the extrinsic content of the picture that makes it a picture? What determines the perceptual recognition of the picture's

---

[1]Lopes [1], Kulvicki [2].
[2]See Wollheim [3].

(external) content? The issue isn't representation in general, but in relation to the effective source of visual vagueness, the joint roles of perception and categorization.

How we recognize the extrinsic content is a complex story, but reports from the psychology of ordinary perception and from the methodology of fuzzy set modeling share the emphasis on the (partial) role of perceived similarities. It is irrelevant whether the similarities are objective as a point of perceptual activity, object features, or the general and relational case I favor that encompasses them and links them too.

Perceived similarities relate equally perceived items, perceived representational properties—visible intrinsic properties of pictorial marks—and visible properties of represented particulars. Both particulars and their properties may constitute the subject matter of the depiction. Regardless, it is possible that the relevant conceptual response might be less the outcome of a judgment of resemblance than of the contextual exercise of a recognitional capacity in requisite conditions. Either case is still one that yields or rests on categorization, carried out by whatever means. Here I concentrate on categorization as the means to examine the extension of fuzziness from linguistic predication to pictorial representation.

From this standpoint, pictorial representations connect four kinds of representational contents: content of seeing, content of seeing-in, content of seeing-as and all forms of excess, incidental, non-pictorial content (visible and invisible). The last three are forms of extrinsic content; it's a metaphysical matter whether the last is also part of the external content, instantiating the features associated with the extrinsic categories that play a role in seeing-as.

The content of seeing corresponds to what I have introduced as the perceptual extrinsic content and pictorial intrinsic content. Seeing-in emphasizes awareness of the supporting medium, the picture as an object of perception, with its intrinsic perceptual features or content; whereas seeing-as emphasizes the extrinsic content. Seeing-as may be indistinguishable from seeing-in, since their associated contents will overlap, namely, when what we see in the picture we may also be what the pictures is seen as. How intrinsic content may be part of seeing-in the question of the difference between seeing-in and seeing-as and arguably the intrinsic content might not be a simultaneous part of the phenomenological content.

In accounts or situations such as pretense, the two kinds of contents collapse into one of illusion. In general, a more plausible position is that the intrinsic content would appear as a diachronic moment in the complex experience of seeing-in. Finally there is the additional extrinsic content of seeing, which includes the target system and alternative pictures of it (as a standard, or a set of standards in the role of a kind). This plays a role in establishing a particular relation of perceptual representation whether in terms of recognition or similarity (besides an indexical causal link or reference, etc.).

Not all content consists in the kinds of categories or associated properties we might consider visible or that are responsible for the visibility of something. In fact, what counts as visible, as already indicated, is a contextual and relational matter of intellectual and material practices, skills and standards.

We are familiar with different ordinary and scientific kinds of extensions of the accepted domain of the visible. Some involve additions to the phenomenological range by means of technological interventions that visualize events, properties or entities otherwise considered invisible: seeing through a telescope or a microscope or any physical detector linked to a visual display. Another strategy involves a conceptual extension by means of inferences: when we see a sign or a part or anything appropriately related to the perceived system, where relations can vary: we claim see an organ on an MRI image that visualizes the outcome of statistical computations, or we claim we see a medical or social condition because we see a symptom causally explained by the condition or phenomenon. All kinds of data are claimed to represent, visually, phenomena that can be identified or understood only with the aid of empirical hypotheses. The two strategies are connected, since some conceptual extension involves inferences grounded on the very kinds of material interactions or correlations that justify the technological extension.[3] These are modes of *content development*, and the information on which they are based is encoded in IC–EC links.

Incidental or excess content may be established explicitly or negatively, by any unintended and unperceived correspondence between any property of a picture, or associated categorization, and any member of an indefinitely large set of items: unrecognized, inaccessible or even nonexistent entities or properties that nominally qualify as external content by virtue of satisfying some minimal requisite condition on any possible property or categorization of the picture or combinations thereof. Nominally, vague categorization can apply to any such relata and relations.[4]

A more restrictive perception-centered account that emphasizes both categorization and resemblance is Hopkins'.[5] On his view, representation requires co-instantiation of visible properties by the representation and its assumed object; the criterion assumes transparency, that is, shared visible properties.[6] An additional restriction requires that the co-categorized objects instantiate properties be not just visible, but be actually seen.

The view is directly linked to the role of categorization, whether or not it results from an opaque process of recognition. According to Hopkins, the content of pictorial representation may be a particular P or a type/property F. Then, to represent a property F is to represent a particular F-categorized in the class of F-things, the base class represented as occupied. A pictorial representation and the thing represented, adds Hopkins, must be in at least one same class F, where F stands for a visible property or a perceived similarity determining membership in F. This is additional to the condition of accuracy of extrinsic content and its overlap with the

---

[3]The significance of this issue in the history of science and philosophy of science derives from the methodological and cognitive significance of any distinction between observation and theory.

[4]For a metaphysical discussion of such entities as intentional objects, see Zalta [4]; for a consideration of their inclusion in the scope of perceptual representation, see Falguera and Peleteiro [5].

[5]Hopkins [6].

[6]This view is more restrictive than Goodman's account of depiction or pictorial denotation.

perceptual categorization associated with the intrinsic content. In fact, this is a common mechanism and standard of perceptual representation, whether or not we adopt it as a general criterion in Hopkins' sense. The similarity may be recognized by description, associated with the relevant categorization, or else by perceptual acquaintance, whether involving the recognition of objects *simpliciter* or their recognition under the relevant categorization, along with awareness of the corresponding instantiated properties.

Representation may be a matter of criterion and a matter of proof, a matter of construction and a matter of calculation, a matter of actual relation and a matter of justification. In most cases—and in some theories in all—the role of inference or evidence is arguably part of the process of construction or application of the criterion of categorization. In the perceptual case, as in more theoretical, conceptually-laden scenarios, the difference between the conceptual criterion and the epistemic testing of its application is stricter than is in other views. The standard is one and the same and in Hopkins's criterion it is perceptual. Representation assumes access to some categorization of the target system P. In the case of unobservable or fictional target systems, the assumption that representation in the picture preserves some structure of represented properties assumes that the target is independently and similarly categorized as well (this is an issue of visualization, not the issue of realism, which would beg the question). In the case of perceived resemblance, the perception of the resemblance requires, by hypothesis and in practice, some perceptual access to the object P or its perceptual representation; some form of shared perceptual categorization or exemplification is the required content of consideration. This is a restriction on the application of IC–EC links for content development.

The class in which we locate the represented object P works as a representation of elements denoted by 'F' that at some level may be considered equivalent or alternatively, a kind, a type and a property that we also claim P exemplifies. This representation, that I have generally called an instance of categorization, might effectively work for depiction purposes through a set of images to which we claim our picture is related. This might be a set of pictures that establish a category of being a picture of P, its individual concept, and it might include one prior public picture of P as a single member of this class and as default standard, or it might be a single memory state from a prior perceptual representation of P. Besides providing the individual concept, the set can also provide the means for the categorization to represent a kind or property, and this in one of two ways, by prior consideration of individual exemplification by Ps or else extensionally, by taxonomical definition in terms of a collection of Ps. The collection of pictures and their recollected perceptions help categorize unobservable and fictional entities as well.

The different modes of categorization of the object of visual representation play a role both in forming the extrinsic content of the picture and in its independent representation acting as standard. How we establish the co-categorization will depend on how the object P is categorized as F—having the property F or being of the F-type or kind—in relation to the F-set: by resemblance to one particular—the grounds for judgments of instantiation—, or to more than one as members of a larger set—that is, joint instantiation and membership are functionally equivalent criteria.

Indeed, co-categorization and co-instantiation of perceptual properties enable representation, although only within a context that includes the following: past experiences, learned habits, recognitional dispositions, conventions, or a knowledge base of background beliefs and information that include empirical associations and theoretical assumptions. Within this context the image producer and the user can recognize features of the extrinsic content beyond the categorization and instantiation that establish the visible intrinsic content.

Now, not all categorization constituting intrinsic and extrinsic content involves co-instantiation. In this case we say that categorization is not transparent. Much recognition involves a holistic framework of internal relations between intrinsic properties that helps identify, in the manner of solving a jigsaw puzzle, more complex categorizations, especially ones that we associate, within a context of conventions, etc., with the extrinsic content or subject matter that is familiar to us from its perception in the world. Examples include the very quality of depth or solidity that perspective and shading simulate on two dimensions, animal signaling behavior may induce defensive, aggressive or cooperative behavior, and the color-coding that in maps, charts or functional MRIs, helps visualize non-visual properties, that is, in opaque ways. Just like the case of spatial intrinsic properties and their internal relations, diagrams and tables visualize non-geometrical, invisible structural relations between quantities or qualities. Opaqueness may be perceptual as well as conceptual.

In light of the perceptual restrictions on the criterion of representation and of the possibility of developing extended content through IC–EC links, we may still try to extend the scope of similarity, and thereby representation, by developing the intrinsic content. Every extension, with an associated categorization, extends the scope of possible vagueness and fuzziness.

Some intrinsic content is perceptual but, in a strong sense, is not pictorial (in a weaker sense of pictorial I use here, the perceptual properties are distinctive of pictures in contrast to linguistic or symbolic representations); and non-pictorial images, while visual, may not depict by direct visual resemblance and co-categorization. Cutting across the semiotic divide between words and images, Peirce called both kinds of images iconic signs, but he called images only pictures, the other icons he called diagrams. Resemblance and co-categorization involving images (icons) may not be established just pictorially, by means of visual transparency alone.[7] The visual properties are said, instead, to present the qualities they cannot represent transparently.[8]

Beyond the scope of perceptual properties they manifest and represent, intrinsic properties are mimetic. This feature of pictures, unlike most images in perceptions – although not all so-called mental images—, is a virtue, an enhanced power of

---

[7]Peirce [7], Goodman [8], Lopes [1].
[8]Kulvicki [2, 9].

representation that extends to the case of representation of fictions and unobservable systems or properties, also in science, where the virtue is of special epistemic, methodological and communicational values between image-makers and users.

One way to think about this fact is to assume that intrinsic content is extended by a number of relational and contextual properties and their corresponding categorizations. This should not be considered anomalous since, as I have insisted, even perceptual qualities are equally relational and contextual.[9] Those properties rest, as does any information to be identified through them, on a number of factors: conditions of perceptual interaction, perspectival relations, background beliefs and interests, depiction systems, recognitional habits and dispositions; the latter are related to one's adopted or familiar standards of realistic representation and, therefore, connect intrinsic and extrinsic contents, categories or information.[10]

Even if we admit a perceptual/non-perceptual or conceptual/non-conceptual dichotomy, the intrinsic content of pictorial representation may be based on some form of identification or recognition, likely a relational cognitive property with a role for capacities and conventions. Non-conceptual recognition might be the cognitive result of a match between the perception state and a memory state of prior perception. Conceptual recognition involves a judgment of perceived similarity or similarity relations within a class or an inference from such relations and auxiliary conventions. But complex conceptual categorizations are generated and implemented differently within specific contexts that privilege, for instance, the following: particular intrinsic properties and resemblances (salience of extracted content) and particular codes of preferred symbolic meanings (often identified in arrangements called systems, styles or formats), adopted by direct convention or symbolic representation of background theoretical and empirical beliefs.[11]

The extended content can accommodate two of the dichotomies mentioned above: object/property and particular/kind. Categorization plays a role in the depiction of elements in either pair. Recognizing or describing individuals isn't separable from representing properties or kinds since the individual is represented as having properties, or connectedly, instantiating a type and being member of a class. In such cases, instantiation occurs in intrinsic properties of an image as well as of the object described in the extrinsic content. Along the generic/specific or abstract/concrete dimension, the categorization and representation may take place at many different levels of abstraction. In science this is the issue of modeling and idealization. Salience of the selected levels in extrinsic content depends on all sorts of methodological, theoretical and cognitive aims and constraints.

Again, part of the recognition or categorization process is the recognition— tacitly or by explicit association—of specific indicators, recognizable signs that in

---

[9]Siegel [10].

[10]See also Lopes [1].

[11]Beyond the scope of resemblance characterizing the role of iconic visual signs, Peirce referred to signs causally traceable or inferable to their content as indices, and to the more arbitrary kind of conventions, symbolic signs.

the cognitive context meet two conditions: (1) they are connected to the type or the classification schema or task—even when clustering and labeling—, and (2) they are properties in the intrinsic content recognized as properties of the image in question. The cognitive task extends equally to identifying an individual person, someone's social or institutional standing, someone's ethnicity, a chemical compound, a variety of plant, a medical condition, a mythological animal, etc.— although contexts, procedures or mechanisms might differ. In the case of individual concepts one might worry that the category is absent, while in the case of fictions, the worry is that individuals are absent. In each case, the identities of the particular real or fictional individual and of the fictional type are associated with a class of images.[12]

In all such cases categorization is at work, enabling the co-instantiation of properties they represent through image-making. Resemblance might play a role with members of the class of images that support the identification of the category, whether individual, fiction or type. Talk of preservation of a structure—a pattern or complex set of relations—is just a way of talking about sharing a feature, co-categorization or resemblance in terms of complex pattern of internal properties —even when the isomorphism expresses a correlation, e.g., diagrams picturing the co-variation of temperature and time, or the correlation between color differences or the height of a mercury column and degree of temperature. In no way, I reiterate, does this role of categorization imply that it is either inevitable or exclusive in effecting representation for viewers or image-users.

The last issue, which I have raised in a previous chapter, is the challenge of providing a measure of similarity. This will depend on specific criteria. But when we talk about degrees of similarity, we must acknowledge a difference between the case of an individual feature and the aggregative case of a set of features associated with a picture and a target system. Here visual representation along the terms examined admits of considerations of approximation and makes sense of the notion of visual approximation. I revisit this issue with more generality in Part 2.

Pictorial representation, then, can be also a matter of categorization, as can be other uses of pictures. The consequence is significant to bring out the differences from linguistic representation I noted above. Unlike the linguistic case, pictorial representation is not a matter of truth in the semantic sense associated with propositional content, at least not in any direct way that involves two of the three steps: (1) analysis into multiple categorizations (intrinsic, extrinsic and external), (2) application of a weaker sense of correspondence between a picture and its external target (in type or token) that accommodates pictorial accuracy or (3) generate a linguistic transcription in terms of linguistic reports of the different categorizations and/or the overall resulting pictorial representation, subject to different kinds of indeterminacy in the absence or presence of the image in question (leaving aside other factors responsible in a given context for the representation function, intended, communicated or used).

---

[12]Goodman [8], Hopkins [6].

Moreover, the role of categorization, unlike linguistic predication, extends to the picture itself, categorized as intrinsic content, as well as the represented object, which is categorized twice, in the extrinsic content and in the independent representation as external target content. Vagueness in representation, then, is a matter in all its dimensions, a matter of vagueness in categorization. And, as additional results, it will be easy to see how the same contrasts with the linguistic case will obtain.

References

1. Lopes, D. (1996). *Understanding pictures*. Oxford: Oxford University Press.
2. Kulvicki, J. (2014). *Images*. New York: Routledge.
3. Wollheim, R. (1980). *Art and its objects: With six supplementary essays*. Cambridge: Cambridge University Press.
4. Zalta, E. (1988). *Intensional logic and the metaphysics of intentionality*. Cambridge, MA: MIT Press.
5. Falguera, J. L., & Peleteiro, S. (2014). *'Percepción y justificación, legitimación o sustento?'*, *ESTYLF 2014* (pp. 441–446). Zaragoza: Universidad de Zaragoza, Libro de Actas.
6. Hopkins, Ch. (1998). *Picture, image, and experience*. Cambridge: Cambridge University Press.
7. Peirce, B. S. (1868). On a new list of categories. *Proceedings of the American Academy of Arts and Sciences, 7*, 287–298.
8. Goodman, N. (1976). *Languages of art*. Indianapolis: Hackett.
9. Kulvicki, J. (2006). *On images*. Oxford: Clarendon Press.
10. Siegel, S. (2010). *The contents of visual experience*. New York: Oxford University Press.

Chapter 10
Vague Pictures as Vague Representations and Representing

In the previous chapters, the discussion of pictorial representation has aimed to motivate different roles and forms of categorization; they include the guiding role of IC–EC links to develop content. The purpose of the analysis of different practices has been to identify potential sites and conditions for the occurrence of pictorial fuzziness beyond the terms set by the linguistic standard. Pictorial fuzziness as indeterminacy or uncertainty can be characterized, to adopt the linguistic standard for vagueness, as the property of lacking discrimination between any possible determinate contents. These are dimensions of our linguistic and cognitive practices, challenging and informing categorization in all their applications. In this chapter I pursue further how different dimensions of representation accommodate the possibility of fuzziness in relation to the specific characterizations of vagueness $V_1(l)$–$V_4(l)$ listed in Chap. 2. These criteria underdetermine their possible semantic content; their generality, I suggest, has consequences for the interpretation of pictures and their fuzziness. Again, my analysis keeps testing the limits of strict application of the standard of linguistic practices. In this chapter I introduce further elements of analysis with additional consequences for the case of images and, as I will note in subsequent chapters, for the interpretation of scientific approaches to pictorial fuzziness, especially fuzzy set formalism.

Remember, this is no mere subjectivism or epistemicism; my view is that the combination of volitional and cognitive subjective elements have minimal objective aspects, as practices relating properties of the linguistic agent and the putatively depicted system. The objectivity is twofold, as cognitive practices that are object of empirical and formal study and as practices presenting an objective, world-oriented, centered pole of a situated cognitive relation that I call *sobjective*; and this minimal objectivity—sobjectivism—is independent of an ulterior ontic commitment to corresponding objective properties.

The pictorial version of $V(l)$ is $V(p)$: What a picture describes is indeterminate, and this concept can be operationalized with the minimal criterion that we are undecided among multiple models or contents; they are not univocally decidable. One example is the indeterminacy of the meaning of realistic paintings, including

© Springer International Publishing AG 2017
J. Cat, *Fuzzy Pictures as Philosophical Problem and Scientific Practice*,
Studies in Fuzziness and Soft Computing 348,
DOI 10.1007/978-3-319-47190-7_10

the distinction between intended and unintended subject matter, or between the object (the model) and subject (whatever else beyond the model that the representation of the model helps represent, e.g., a historical figure or fictional character).[1] This situation prompts the standard criticism of similarity as a criterion of representation, since similarity only multiplies, not to mention by degrees, the number of actual and possible subjects.[2] The picture is indeterminate about what it shows, about how it should be categorized as a whole or any of its parts, about what should be recognized in it. It is a problem of indeterminate perceptual content.

In the non-human natural world, something like indeterminacy of reference may be said to extend to visual features of animals or plants that, in richly constrained contexts, operate as effective signals, that is, as visual stimuli that may be modeled as a form of communication. To fit my approach, any minimal senses of representation and its content will have to be reduceable to notions of whatever process plays the role of categorization.

In the linguistic context, the possibility of vagueness can be objectivized in the form of a property of the subject of predication and of a true statement, as modeled by fuzzy set theory.[3] Now we can attempt to establish a working notion of pictorial vagueness by means of V(p) criteria. It will be important that the criteria apply to intrinsic and extrinsic properties of the representation and to any related properties, perceptual or theoretical, in the representation of the target system.

In this chapter I focus on criteria V_1—borderline status—and V_4—closeness; they are the ones that best facilitate the application of mathematical modeling to fuzziness of pictorial categorization. We may distinguish at least six different cases according to the kind of subject matter attributed to an image (as I have observed, they are hardly separable): (1) Intended and declared particular, (2) intended and undeclared particular, (3) unintended particular, (4) intended and declared type, (5) intended and undeclared type, and (6) unintended type [(3) and (6) fall under what I have called incidental content].

The relevance of subject-types spans over a variety of pictorial uses, from education to scientific practice. In the sciences, priority is typically given to the intended portrayal of kinds, types or properties—whether directly or indirectly—instantiated by a properly categorized individual. Individuals of interest are identified derivatively by coordinating different generic descriptions.

The challenge now is to show how putatively vague categorization of the target system can satisfy vagueness criteria V_1 and V_4. Recall that categorization plays a role in determining the intrinsic and extrinsic contents of the representation and, where relevant, also of the independent representation of the target system (perceptual or not).

---

[1]The subject of representation, which I sometimes call object, should not be confused with the subjective element of the cognitive activity; considerations of subjectivity apply only to the latter.

[2]The example of Rorschach test might be relevant, especially when the emphasis is placed on the period of individual hesitation or the possibility of experiencing Gestalt switches.

[3]This characterization covers Zadeh's original empirical project in Zadeh [1] and Smith's more recent philosophical motivation and explication in Smith [2].

On criterion $V_1(p)$, the interpretation of vague pictures would yield borderline cases. The application of $V_1(p)$ is a matter of borderline categorization of two things, the picture (as a particular or, in the case of a type of picture, a group of similar enough particulars) and the object of representation (the property or kind of a thing, broadly conceived to include facts, events, states, etc. or of a quality). The represented object figures in both the extrinsic content and in the independent external representation. An object of categorization is a borderline case if in our meaning-fixing practices we, scientists and non-scientists, individually or communally, cannot agree or even judge whether it falls under a specific category—or instantiates a property, or belongs in a class. Of course, a precise definition of borderline case is part of the problem; only a working characterization, sharpened for the sake of applicability, will fence off second-order vagueness; it is not an adequate alternative to reduce it a priori to an epistemic matter of ignorance.[4] As a matter of categorization, not just competence in linguistic practices, we might find nevertheless that an epistemic dimension also features in our determination of borderline cases—a fact my generalized perspective can accommodate. The second criterion, $V_2(p)$, requires predicates having an extension with a blurred boundary. The criterion reduces to the assumption that borderline cases cannot be eliminated.

The first potential borderline object is the material picture itself; the borderline criterion must apply to some categories, however determined, standing for an intrinsic property of the picture, or a similarity class, or set. As I have argued, the most typical of relevant intrinsic properties are geometrical, topological and chromatic. They constitute surface marks, a content that overlaps with the extrinsic content of perception, and is phenomenologically indistinguishable, in the way I have discussed, for instance, for the case of blur and fuzzy photographs: we cannot definitely identify a color, or the tonal contrast as a sharp line further recognizable as an edge, or categorize a mass of color as of some particular kind of geometrical shape.

In different contexts, the interpretation and use of pictures are informed by different cognitive and practical aims and values. Fuzziness is not a purely semantic affair. What counts as a standard of precision is a matter of context, of what counts as precise enough or, alternatively, what kind and how much fuzziness is tolerable or acceptable given a set of constraints. Relative to a given standard of high definition, for instance, it is still common practice in many areas to introduce visual techniques of compensation or correction for undesired fuzziness especially in intrinsic properties of pictures. One strategy I have already mentioned consists in superimposing more precise diagrammatic lines and colors that help identify, select or emphasize visual elements—circles, arrows, etc. The addition of such visual elements supplements images for the sake of their use and not just aesthetic appreciation; and users have to engage in constrained practices of production and interpretation.

Most often these strategies of enhancing visual precision, or definition, serve ulterior purposes of optimizing the tasks of recognition or interpretation for the sake

---

[4]Smith [2].

of using the information in the extrinsic semantic content. They are guided and constrained by operative aims and standards of, for instance, diagnosis, analysis and communication.

In digital imaging, processing takes the form of computational operations on an initial image such as filters and transforms; among the purposes they serve are enhancement, restoration, compression and segmentation, which serve in turn further cognitive and practical purposes.[5]

Another dimension of intrinsic visual vagueness is compositionality. Recall the Fodor–Sober compositionality condition associated with the distinctive character of images, that parts of pictures represent corresponding parts of the intended object. In general, precision is associated with the capacity to discriminate, and this is a general sort of compositionality condition that applies to quantitative forms of categorization in measurement: distinctions in representation track distinctions in (categorization-relative) features attributed to the target system. I have already pointed to holistic, Gestalt-like effects, typical of paintings and perceptual realism, which emphasize the roles of distance in perception and size and detail in composition. This is the issue of resolution, or the lower limit of validity of the compositionality condition allowed by the molecular structure of the pictorial medium. There the analogical image becomes effectively digital (relative to pictorial function of the parts, not all physical and visual properties, since pixels form contiguous arrangements). There is a limit to our capacity for precise categorization, set by the scale at which pictorial functionality breaks down. But the phenomenon I want to note first is intrinsic, whereas the Fodor–Sober condition isn't. What relates the two is the fact that extrinsic, semantic compositionality is informed by intrinsic, syntactic compositionality.

In digital media, the limit is set by the size of the picture elements, the pixels. The larger-scale effect becomes visible in proportion to the size of the screen and the number of pixels. From the compositional perspective, micro-vagueness becomes a matter of the possibility of vague categorization of pixels as (1) individuals with a range of intrinsic, medium-dependent properties and (2) as collections of indistinguishable parts. Notice here that the compositionality of visual intrinsic features is a holistic affair. Pixels contribute with their properties to properties of the image as a whole or at a larger-scale that are not their own, such as geometrical features of size, curvature and shape; these are the product of tonal properties such a levels of brightness that combine also into larger-scale chromatic effects.

In fact, relative to certain cognitive and practical purposes, higher-level and larger-scale precision, intrinsic and extrinsic, may depend on violating the condition of compositionality. The whole is rendered more precise at the expense of the precision of the parts. More generic categorizations leave indeterminate more specific details that would otherwise make the picture richer, more realistic, fleshed out, determinate. Visual and conceptual modeling practices are examples. Schematic, diagrammatic images are of this sort. We identify fewer details, or if we could, in the

---

[5]Castleman [3].

context at hand they aren't relevant. Judgments of emphasis and simplicity inform the construction and evaluation of pictures. Image processing often needs to incorporate such judgments. In other cases, despite intrinsic richness, we cannot really fully apply the extended categorizations, we can't quite unequivocally identify something in the picture, whether intrinsically fuzzy or not; the vagueness is extrinsic. We can't quite identify the person's gender or ethnicity, or the structure of something in a microscopic or telescopic image. The challenge defies the application of taxonomies or categorization schemes; it is not a matter of accuracy.

To the extent that all such intrinsic properties are perceptual, they are relational, especially interactive—and this relational aspect doesn't conflict with being amenable to objective understanding and representation, formal and informal. Although the brightness level of the pixel might not be individually discriminated, and in electronic media the intrinsic properties are categorized in physical and mathematical terms such as intensity and frequency of light. The choice of quantitative concepts is not just part of how the physical processes of production of images are understood and applied; they enable the computational means for processing images by controlling the individual and collective properties of pixels in light of preset cognitive and aesthetic goals and standards. In many kinds of practices, the reduction of fuzziness is a common goal.

$V_1(p)$ applies to the categories informing extrinsic content as well, what the picture represents, what it shows or what we see in it. The difference lies in the extrinsic subject of categorization and the range of relevant categories. The range provides the basis for determining objects as well as aspects, for applying a distinction between object and property or object and subject matter, referent and description. The two are inseparable, since the categorization of objects often relies on the categorization of features, and (recognized) reference often relies on description. How the distinction is applied will depend on each context, for instance, after we use the pre-established concept (concept cluster) for an individual entity or type of entity—theoretical, fictional or considered actual—we use additional categorization to qualify it; this is how we attribute properties to planets or fictional characters.

The range also enforces, contextually, distinctions between perceptual and other descriptive categories, simpler and more complex conceptual structures. Visualization of non-visual properties and transparency of shared visual ones extend what counts as extrinsic content. The extension is based on background associations to the shared intrinsic properties; this semantic link may take many forms: symbolic, conventional, theoretical or empirical, an inference rule. Some of them may be applied even through auxiliary linguistic terms or statements: paintings' titles, names on maps or text figures' captions.

For instance, medical symptoms or any indicators are standard cases; the association is the indexical representation by trace and correlation: certain clouds represent rain, certain shapes on X-ray images represent some type of cancer, certain curves in financial graphs trigger warnings, certain spikes in energy data signal the presence of a particle, certain facial expression represents an emotional state, certain color represents a threat level on a chart, and so on. The case of visual representation in measurement of invisible properties is just a quantitative version. Representation and reasoning, picture and evidence, are thus hardly separable. Recall my discussion of data, visual and visualized, above. In the sciences, this is a classic general issue in empirical conceptualization and testing. As matter of ordinary cognition, the scope of extensions of the inferential kind (by means of transparent rules or empirical associations) and non-inferential kind (opaque processes) are even more general. Either way, in the extension, the coordination must be fixed conceptually, although it gets empirical support from interventions and techniques.

We are led back to the original matter of interpretation. Is vagueness a matter of semantic indeterminacy or an objective property of the object of categorization? If the categorization is a relation to some property, it must be a relation of partial instantiation; then vagueness cannot be a semantic relation except as a matter of degree of truth or partial membership in a model set. This instantiation may involve a relation of degree of similarity to another particular instance, namely, a standard or prototype for the purpose of categorization. As a consequence, the borderline character of the object of representation might be assessed (not defined) as a matter of dissimilarity from the particular that counts as the standard, which can be the picture (rather than a third system). This degree of dissimilarity would explain the distinguishing mark of borderline cases, our failure to decide if the picture represents the designated object. It would also provide a structure for applying a notion of partial membership in a set standing for the category in question.

Then there is the categorization of the object of representation, effectively an independent representation, perceptual or pictorial. A picture is effectively like an actor performing a role. It bears two contents, two overlapping products and processes of categorization. There are at least two ways to consider the matter of accuracy of the extrinsic content. From a semantic standpoint, it is a factual matter about whether the object instantiates the properties presented in the content (in all the ways listed above). I have called this the condition of co-instantiation. From a cognitive or epistemic standpoint, it is a matter of acquaintance with that fact, a matter of evidence about the degree of co-categorization and matching. Aside from vagueness in the independent categorization of the object, vagueness enters the picture to the extent that the relation seems indeterminate relative to the independent base categorization; it is not decidable whether the picture captures the actual object that as pictorial actor it is impersonating. We may declare it a demonstrative kind of fuzziness, indeterminacy of categorization-dependent ostension.

The last criterion I consider is the closeness condition, $V_4(l)$: that F is vague implies that for any two objects, statements that they are F are close in respect to truth. After $V_1(l)$, one may consider two pictorial versions. The first might read like this: if a picture is vague, for any two objects, the vague picture is a picture of one and is very close in pictorial "truth" to a picture of the other, where 'very' is fixed for a given context. In terms of categorization, in its role of representation instead of predication, it requires that the categorization of one object be very close to that of another; for intrinsic content closeness involves another picture; for extrinsic content, another object of depiction. The second, $V^*_4(p)$, might read like this: what a picture communicates is captured by its categorization in intrinsic or extrinsic content, and it is vague if, applied to two different objects, its categorization leads to two pictures very close in pictorial "truth" content.

From the previous chapter, it must be clear by now that what I'm calling "truth" in the pictorial context can only be a semantically weaker relation that is best characterized, at least as an abstract placeholder, in terms of some notion of correspondence, verisimilitude or a somewhat more specific relation such as representational accuracy that does not share features of truth associated with a propositional expression. For "truth" content we can substitute pictorial content. Content may depend on—or, more weakly, is enabled by—, for instance, conditions of reference and the degree of instantiation of the properties the picture attributes to the object, particular or type.

Semantically speaking, I have suggested, we can rely only on closeness in respect of categorization. But closeness within a set margin is incompatible with the disjunction between two states of determinate co-categorization, indistinguishability and contrast. As a consequence, the suggestion is that categorization be allowed a condition of degree or partiality, just what fuzzy set membership can model, albeit, contrary to the objective linguistic case, without conditions of degree of truth. Substituting the notion of degree of accuracy eliminates any interesting difference between vagueness, or imprecision, and inaccuracy.

For cognitive purposes, more cognitively relevant than truth is whatever function truthfulness might have and how we can establish it; for instance, as a stopping point of inquiry, of testing—whether for warrant in terms of reliability or inferential support, provisional and revisable. This is the basis for an epistemic version of pictorial vagueness. This one does not depend on the standard epistemic interpretation, according to which, there is a line in the conceptual or metaphysical limning of kinds or properties, independently of the further issue of whether concepts or properties are more fundamental; the epistemic subject simply fails to know where the line lies (Smith calls it the location problem). There is no sharp line with a location to be known. It is an indeterminate matter whether some object possesses a property or relation; they neither do nor don't. The situation is independent of the degree of precision presented by the (visual) representation, which in this view is not vague or inaccurate. In the pictorial case, the intended object represented by the picture may be perceived with imprecision, as pictorially vague, with an imprecise visual property, or, equivalently, as imprecise.

Pictures don't just appear vague; they are vague in either of their possible kinds of contents, intrinsic and extrinsic. Again, I am associating vagueness with potential indeterminacy inclusive of the possibility of a degree of objective determinacy. Fuzzy membership is a formal measure admitting of different interpretations, for instance, as a factual property of things (Smith) or as a factual form of categorization behavior (Zadeh).[6]

---

[6]Smith [2], Zadeh [1].

The final question is how the different objective dimensions of determinacy may combine or relate. I have addressed this question in relation to the content of perception, but also in relation to realism and pictorial content in photography and painting. Seeing fuzzy, seeing a fuzzy picture, seeing a picture of a fuzzy thing and seeing a fuzzy thing may be under certain conditions phenomenologically indistinguishable. Even if one distinguishes between situations of perceptual contrast in which fuzziness is peripheral and those in which it is central—with peripheral sharpness—, each class of situations leaves room for indistinguishability between at least two of the kinds of experiences. But more interesting relations may be possible. The results may be formulated in terms of the violation of two specific forms of the general IC–EC link, (1) the (PIC) precise intrinsic content–(PEC) precise extrinsic content link and (2) (IIC) imprecise intrinsic content–(IEC) imprecise extrinsic content link.

Cognitively and aesthetically, photographic realism became the endorsement of intrinsic precision as a norm of pictorial practice—I call it material realism. It is based on the perceptual and theoretical assumptions about target systems, that they can be understood and perceived sharply, and relative to such a standard we may assume a "measure" of pictorial accuracy or "truth". The standard leaves fuzziness as an aesthetic or practical choice. A sharp-looking picture, with precise intrinsic content, may indeed have an indeterminate—or vague—extrinsic content, as well as multiply determinate ones. The indeterminacy can be traced to a number of contributors to the job of representation such as intention, pictorial conventions, information, recognition capacities and competence or causal indexicality. In such cases, when we establish that certain purposes require a higher degree of determinacy of intention and interpretation, we rely on supplementary information presented in linguistic form. The indeterminacy may be reduced by re-interpretation as an objective property—relational or contextual—of the system in question. But as a matter of linguistic truth, the "interaction" hardly seems to guarantee equivalence, the linguistic description does not transcribe and replace pictorial content.

Alternatively, fuzzy-looking pictures may be realistic too, with "accurate" and "determinate" extrinsic meanings. At best, the relation between vague descriptions and "vague" systems can only be understood, as in the case of material realism, a relation between two categorizations; but the two situations are not characterized by the same relation. The operative notion of realistic adequacy of vague depiction involves a more abstract, higher-order notion of accuracy and determinateness. This epistemic standard is based on a dual standard of accuracy, a perceptual standard for the representation of a fuzzy entity based on perceptual beliefs, and a supporting or separate theoretical one based on a (theoretical) model or conceptualization of the entity that doesn't postulate, for instance, sharp spatial boundaries. Of the perceptual kind, I have mentioned the pictorial commitment to fuzziness as standard of perceptual cues for distance, volume, texture, motion and focus of attention. This challenges the general formulation or systematic application of a PIC–PEC rule, or any specific fixed form of the IC–EC link.

Based on the terms of the standard $V_4{}^*(p)$, we can suggest two additional semantic pairing (meta)rules that accommodate different IC–EC links: precision-accuracy and imprecision-inaccuracy. Cases that challenge the interpretive validity of the PIC–PEC and IIC–IEC links or rules also challenge the meta-rules associating precision with the expectation of inaccuracy. Accuracy of perceptual or material representation often follows the perception of fuzziness. Fuzziness enforces exactness of categorization as well as its precision. We can add, then, two alternative semantic rules: precision-inaccuracy and imprecision-accuracy.

References

1. Zadeh, L. A. (1965). Fuzzy sets. *Information and Control, 201*, 240–256.
2. Smith, N. J. J. (2010). *Vagueness and degrees of truth*. Oxford: Oxford University Press.
3. Castleman, K. R. (1996). *Digital image processing*. Englewood Cliffs, NJ: Prentice Hall.

Chapter 11
The Cognitive Values of Imprecision: Towards a Scientific Epistemology, Aesthetics and Pragmatics of Fuzziness; Contextual Lessons from the History of Picture-Making Practices

Vagueness, as imprecision or fuzziness, concerns many practices and many values. Epistemic and aesthetic practices are in the case of images, as well as of words, inseparable, e.g., in relation to the value of depiction or representation. Then out of such common ground we may easily see that the goals or norms that inform the uses of production, interpretation and uses of images might be equally entangled yet still differ. In the cases I turn to in this section those values and standards involving the treatment of visual fuzziness are made explicit, and aesthetic discussions yield epistemic insights about perceptual cognition.

Art and cognition, picture-making and perception, are not just inseparable but understood in terms of scientific practices and standards. We may refer to scientific epistemology and aesthetics of fuzziness from that standpoint, and not just in relation to their concern with fuzzy images with scientific value. The formal and technological application of fuzzy-set formalism is an example of a scientific epistemology and even aesthetics.

Aesthetic considerations of fuzziness aren't new. Conceptual ideals of sharpness, notably with the rationalist roots, have long informed epistemic standards in relation to definition and clarity of intellectual representation. The abstract conceptual fixity contrasts with the dynamical connotation of vagueness (from the Latin *vagus*, roaming), with negative epistemic and moral implications (e.g., mistrust of vagrants and wanderers). The tradition of normative precision has adopted a binary distinction between sharp and fuzzy. Similar standards have been applied to sort out

© Springer International Publishing AG 2017
J. Cat, *Fuzzy Pictures as Philosophical Problem and Scientific Practice*,
Studies in Fuzziness and Soft Computing 348,
DOI 10.1007/978-3-319-47190-7_11

sensory representations or, like Descartes, to mistrust them in kind as a source of knowledge –despite the optical analogies underpinning the notions of clarity and sharpness of focus.[1] The precision standard did not deny value to fuzziness, but it restricted it to its relation to the aesthetic exercise of the productive imagination.[2] But in relation to cognitive lessons and assumptions in artistic representation, I will discuss a few cases that are connected to the empirical sciences.

I begin with the practices involving photographic images. I believe the case for the determinate extrinsic content of fuzzy images dates back to the mid-19th century, in the footsteps of the introduction of photography. It is not the experience but the issue and its significance that I am trying to document. These cases are instructive because they suggest that precision emerged as a visual value alongside its quantitative counterpart. Debates were fought over the status of photographic images and the question of which discipline, science or fine art, and which corresponding value, scientific truth or artistic effect, the standard of definition was meant to serve. Disagreement arose over the meaning of realism, mechanical and artistic, and along with it the role of clearness, sharpness or exactness of definition as its standard. The opposite of precision is not homogeneous in either form or meaning. Professional and practical efforts were directed to establishing a catalogue of flaws and their equally diverse etiologies and forms of prevention.[3]

Two main positions can be identified. They made reference to imprecision and equivalent concepts such as, haziness, blurriness, fuzziness, unsharpness, etc., in order to express a failure of definition of form and variation of tone. On one view imprecision was associated with artistic ideals of subjective feeling and beauty, and precision with truth. This is the view that elevated the standard of photographic realism in the form expressed by a family of virtues such as contrast, detail and resolution associated with the mechanical ideal of, in Joseph Nicéphore Niepce's term, 'automated reproduction'.[4] Truthfulness was associated with exactness, identified as 'the tendency of the age' and, more socially, the emerging distinctive mark of professional skill.[5] It is by the scientific standard of ideal exactness that photography sought recognition and service to science.

On the other view, exactness could conflict with truth and hence could not be identified as its reliable standard. Amid the mid-Victorian debates over the boundaries between science and art and the status of photography as a technique

---

[1]The problem of the relation of experience to knowledge has often been framed as a problem about the relation between precise cognition by construction, after the model of mathematics, and imprecise experience in intuition; notable examples include works such as Moritz Schlick's *General Theory of Knowledge.*

[2]I mentioned the case of music in Chap. 1, above. For a general discussion of the aesthetics of fuzzy images and its contemporary artistic endorsement see Huppauf [1].

[3]Hentschel [2]. On the idea of accuracy as an epistemic virtue see Williams [18]; on precision as a 18th and 19th-century scientific and moral values linked to ideals of reason, objectivity and truth, see Wise [3].

[4]Niepce [4], 39.

[5]Niepce [4], Cat [5].

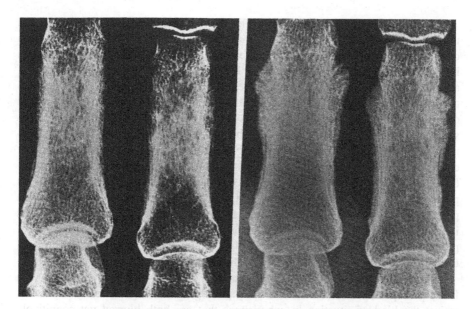

Fig. 11.1 X-rays with high contrast (*left*) and soft contrast (*right*), from G. Spiegler's *Physikalische Grundlagen der Röntgendiagnostik*, 58

and a kind of picture, a novel episode in the history of the values of precision emerged. In it, the second view challenged the identity of exactness and truth that echoes the scientific ideal of mechanical objectivity of which mathematics was a paradigm. Instead, less scientific commentators echoed the sentiment of Elizabeth Eastlake's sensitive defense of 'true gradation' against strong contrasts and 'high lights' that are 'far from being true to Nature', H.P. Robinson's warning that 'the single result of [mechanical] realism is falsity' or P.H. Emerson's cautionary advice, 'Do not mistake sharpness for truth.'[6]

A sharp contrast, then, may be an artificial standard that fails to record the visible and to render visible the invisible. The blinding edge becomes an inadequate standard of representation that can be found also in the scientific uses of photography. In medical diagnosis, X-ray photography has a long history of cautioning against the suppression of information about surface structures such as the outline of bones with pathological conditions (Fig. 11.1).[7]

How is this debate relevant to my discussion of the content and properties of visual perception? In support of the representational value of imprecision, we find in the 19th century a modern defense of two forms of realism, I call them perceptual realism and material realism, respectively; the differ in their standards of realism, the former's is cognitive, the latter's is ontic.

---

[6]Eastlake [6], Robinson [7], Chap. 9 and Emerson [8], Chap. 4.

[7]See, for instance, Gottfried Spiegler's *Physikalische Grundlagen der Röntgendiagnostik* (Thieme 1957), cited in Gombrich [9], 183–184.

The visual content of cognition provides the standard of perceptual realism. This is the canon entrenched in artistic pictorial tradition since the introduction of rules of perspective (the two-dimensional representation of a three-dimensional world as if seen outside a window). Gombrich, the historian, psychologist and philosopher of art, has called this standard the subjective standard of truth, as perception involves aspects that are uniquely individual. Yet they provide a pictorial standard; the field of vision, as Gibson noted, is treated as the original picture. The realistic representation of something must bear resemblance to, and by implication, be categorized in the same way as, the way we experience it 'directly'. Gombrich notoriously called such phenomenon of recognition an illusion effect that provides an operational standard of representation, namely, one can easily feel (if not pretend) in the perceptual presence of the visible system as if in the conditions that characterize perception. As I have already noted, they include modes of limitation such as incompleteness, indeterminacy and distortion, as exemplified by the effects of perspective. According to perceptual realism, photographic precision isn't truthful to reality as we experience it, that is, to the contents of perception.

Where does fuzziness enter this standard of realistic picture by analogy with fuzzy perception? As discussants of early photographic images already noted, this is just the kind of situation in experiences that are both blurred yet cognitively informative or aesthetically effective, or both, with gradation of sharpness in focus yet with a determinate extrinsic content: experiences of volume, distance, surface texture, motion and attention inform the interpretation and production of pictures. Sharpness of boundary distinguishes figures from background, where other figures are perceived as background, through a sharp enough contrast in brightness. This is not the only contrast. In the visual arts, including modern color photography, we can also consider chromatic categorization and the perception of color differences.

In his studies in the history of art, Gombrich identified the significance of vagueness in pictorial representation and placed it in a rich context of cognitive psychology and social history.[8] For Gombrich this significance derives from complex conditions of perception shared by picture-makers and viewers.[9] His discussion has two virtues. Besides the scientific grounding of his empirical analysis, some of his insights fit nicely with the categorization approach I have adopted here, without committing to some of the details of his social and cognitive models. Moreover, the generality of the relevant conditions and mechanisms extend their relevance beyond the case of more distinctively artistic pictures, that is, those whose production and appreciation rest distinctively or primarily on aesthetic considerations or take place in so-called artistic situations.

Gombrich's framework borrows related ideas from Gestalt psychology, Gibson's ecological approach, information theory and Popper's dynamics of conjectures and

---

[8]Gombrich [10].

[9]Obviously, among the viewers are the makers themselves, with the rare exception of blind artists. A number of different conditions impairing an artist's eyesight may play a role in the making of pictures driving a wedge between the standards of production and appreciation.

refutations (Popper's is a method of assessing scientific hypotheses, originally a Darwinian model of child learning by trial and error and later also of machine learning). The driving notion is that perception at the basis of depiction and appreciation has a constructive and hypothetical character; the artist extends and externalizes this constructive process through the activity of picture-making. The key aspect is the assumption that schema precede perception, in the mind or on the canvas, and this precedence is constructive—'making before matching.'

Recognition involves what he called projection; I take the idea in a generic sense compatible with a number of different characterizations and explanatory mechanisms. It begins with the recognition of perceptual schema, an approximate category, often simple or ideal. In perception we refer to prototypes from the perception of particular specimens or pictures. In art, especially art training, the starting points are visual models, ranging from designated prototypes to elementary patterns for their representation (for instance, elementary geometrical shapes or colors and a number of their compositions).[10] Notice the similarity with the role in the sciences of paradigmatic model systems and representations (including formal mathematical concepts and their application in theoretical idealizations such as linear models). Then, the activity of visual depiction, the copying or representing, proceeds through what Gombrich called the rhythms of schema and correction. The hand and the eye introduce the particular details of depiction that de-idealize the preset schema to close the pictorial gap and fit the perceived features of particular object.

The value of Gombrich's and also Lopes' accounts resides in the empirical diversity of kinds and sources of such schema they allow.[11] Some patterns are constrained by mechanisms enforced by the viewer's familiar environment; some might have a genetic component such as ones enabling animals to identify their preys or predators; others are informed by limitations set by the characteristics of instruments and the material medium of creation or display; others have pedagogical value as mechanical part of the training process and happen to be simple enough to facilitate it and its subsequent application; others rely on the mechanical character of the process in order to enable the mechanical reproduction of some system for commercial purposes; others such as elementary geometric figures long enjoyed a metaphysical and aesthetic significance dating back from Plato's geometrical cosmology. In a related tradition of natural history, running til the mid-19th century, kinds were represented by more elaborate schemas depicting their ideal member, the archetype.[12] Since Plato's universals and Aristotle's essences, this is just another episode in the history of the epistemology of conceptualization or categorization, from classification to explanation. Then, with the advent of Darwinism and the popularity of

---

[10]The original conception of pattern and model can be traced to these kinds of practices. Outside workshops and art schools, pattern books became popular in the teaching and practice of all sorts of crafts.

[11]Like Popper, Gombrich aimed to naturalize through empirical models the philosophical standard of universal and necessary categories in Kant's constructive metaphysics of empirical knowledge.

[12]On the role of the artist in the depiction of archetypes see Daston and Galison [11].

photography, the problem of representing the concept, the kind or the universal became linked to the representation of visible particular instances.[13]

The Greek geometrical tradition constrained the representation of the natural world in astronomy and mechanics, not just in the disciplines of picture-making. Until the 17th century, natural motion was considered to be circular on account of the alleged geometric perfection of the maximally symmetric closed curve and planetary orbits were assumed to be circular accordingly. Galileo's mechanics and Copernican astronomy illustrate the role of this particular schema. Also Galileo's artistic training in the technique of chiaroscuro to represent regular solids by the shadows they project played a part in his astronomical observations and reasoning; by enabling him to identify the darkness on the moon's surface as evidence for its irregularity and for accuracy of the Copernican system of the heavens.[14]

Projection is an activity that is expressive of the artists' and the viewers' stock of schema and background knowledge more generally. The author's must rely on shared recognitional habits and background knowledge if he expects to communicate a certain conception or effect. He might supplement the pictures, as in scientific illustrations, with additional information in the main text or captions. If projection is key to the viewer's experience, to his interpretation and appreciation of the picture, it is often so in spite of the author's intentions, also beyond the projection of basic perceptual schema. Authors and viewers form communities around modes of visual categorization based on learned conventions and skills and shared pictorial prototypes. In different contexts the viewer will identify the intended subject of the picture, real or fictional, with specific symbolic meanings of appearances, objects, places, signs, colors, etc. Intended and elicited responses may range from the emotional to the more abstract, symbolic recognition, as in the case of scientific observations. Learning to see things and to see things in pictures provides the resources for categorizing and interpreting. Any cognitive model must be supplemented with social models of transmission of the relevant information that educates picture-makers and viewers alike, whether it's impressionist or cubist paintings, weather maps or Feynman diagrams.

I have already noted that, from this standpoint, pictorial description hardly differs from linguistic description. Goodman identified the visual difference between pictures and symbols in terms of analogical and digital systems; instead, Gombrich, and more recently Lopes, have emphasized the conventional, contingent and contextual similarities between the both systems. The shifting standards, skills and habits of recognition and categorization have long informed the practices of depiction and appreciation without relying on the extremes of linguistic conventions—"natural signs"—imitating Nature.

How does fuzziness enter Gombrich's empirical account of pictorial representation? Gombrich seems to hold a view with two different parts. Again, it might help to assume the distinction—admittedly contextual and vague—between two

---

[13]See Daston and Galison [11], Cat [5].
[14]See Kemp [12].

kinds of uncertainty, one concerns the perceptual intrinsic content we associate with some form of blur—when we fail to identify boundaries, shapes or colors—; another concerns more abstract schema of extrinsic categorization, as when we cannot identify the sharp picture of a distant bird, a set of spectroscopic lines or a microscopic structure. The two parts of his account are complementary and apply generally to both types of cognitive—and aesthetic—situations, although his discussion centers on the application of one view mainly to the intrinsic case of perceptual realism and fuzziness in the visual arts.

One part of the psychological account emphasizes the specific structure of the process of schematization. The general dynamics of schema and correction turn the cognitive trial and error into a learning process in which the schema may have any level of generality and abstraction. What gives the process a directional sense of progress is a twofold epistemic focus: the aim of correctly identifying a particular object of experience and the process of correction resulting from the avoidance of uncertainty and error. However, the standard of correction is not directed by the standard of correct representation. In good Darwinian and Popperian fashion, correction is the outcome of refutation of visual guesses and anticipations by contradictory detail that suggests alternative categorizations or conflicts with what we consider established visual background information.

This process of correction has an additional sequential structure analogous to the game of Twenty Questions, where, in Gombrich's words, 'we identify an object through inclusion and exclusion along any network of subclasses (....) to narrow down our concepts by submitting them to the corrective test of "yes" or "no."'[15] He presented the model of pictorial learning not just as a model of human cognition, but also as an early model of machine learning. The sequence of modifications that individualize the categorization is increasingly discriminating and sensitive to particular differences; it is meant in this way to close the gap between elementary schema and the recognition of the individualized representation. The network that informs the process typically follows a tree-like structure that grows through a hierarchy of concepts, from general and abstract to particular and specific. While the simplified or idealized basic schema or pattern may be well-defined, relative to the final picture, Gombrich considered them vague, that is, lacking in determinateness or specificity.[16] It might well provide a generic depiction of the broad concept and universal, but not a specific visible individual. He also called it initial vagueness; its cognitive value resides in its flexibility and contingency, with the pragmatic virtue of allowing the individualizing series of categorizations.

The resulting correctness is approximate and valuable in a pragmatic sense, relative to a purpose. The corresponding degree of vagueness or specificity is meaningful as part of a picture-making process and is valuable also pragmatically.

---

[15]Gombrich [10].

[16]The French philosopher Pierre Duhem made a similar point about the relation between the quantitative representations in measurement, or theoretical facts, and qualitative empirical facts, or practical facts; he concluded that physical laws were neither true not false, only approximate; see Duhem [13].

The representational value of the image, then, rests less in its lifelike similarity than in its practical 'efficacy within a context of action.'[17] Gombrich gave the example of the process of police identification of criminals. The profiling of suspects takes place from basic templates and applies to linguistic questioning and to pictorial sketching: 'They may draw any vague face, a random schema, and let witnesses guide their modifications of selected features simply by saying "yes" or "no" to various suggested standard alterations until the face is sufficiently individualized for a search in the files to be profitable.'[18]

Still, the acts of categorization that make up the identification process set limits to the scope of fuzziness. Each act of categorization has a binary, yes–no, structure associated with the added visual detail; the linguistic proposition embedded in each question is decidable as either true or false (regardless of how we understand truth or truth value). In the context of the image as a whole, each added visual feature and categorization might be a determinate representation relative to the recognized feature of the subject (presumably also in the holistic context of the "remembered" perception). For Gombrich only the initial schema is vague—and potentially also the resulting image.

The second element in Gombrich's psychological account is also general and relies on the basic roles of fuzziness in training, production and interpretation. While the first—based on pattern or schema—introduces the dynamics of making and matching, the second introduces the complementary dynamics of suggestion and projection. The second is based on the suggestive role of the inkblot. According to his general model of projection, the viewer must be given a "screen" onto which he can project an expected image, more specifically, an area that is empty of ill-defined.[19] So, from a general cognitive perspective, categorization may be binary and even more or less abstract; but for Gombrich the mechanism of projection is triggered by visual fuzziness: first in the fuzzy perception of objects in the environment and more specifically in the visual interpretation—and production—of pictures. Its fuzziness may extend over a wide range of contents, but it begins with the intrinsic geometric and chromatic features. In the former case, the value of fuzziness may vary by context. In the latter, the role of visual fuzziness accords with the perceptual standard of realism—the naturalistic style that Gombrich called the subjective standard of truth. Both are expressions of 'the power of indeterminate forms.'[20]

The standard became encoded in techniques of the realist pictorial tradition such as *chiaroscuro* and *sfumato*. The latter was masterfully developed, discussed and applied by Leonardo da Vinci, 'the inventor of the deliberately blurred image.'[21] For instance, with the aid of color, da Vinci championed chromatic variation to add

---

[17]Ibid., 110.

[18]Ibid., 88–9.

[19]Ibid., 208.

[20]Ibid., 219.

[21]Ibid., 221.

to the geometry of perspective a new aerial perspective.[22] A resulting aspect of landscape backdrops is their distinctive misty appearance. This naturalistic illusion depends, as in geometric perspective, on eliminating, selecting and distorting visual information. Da Vinci called it in his *Treatise on Painting* the 'perspective of disappearances', with an order of schema that will disappear—or appear—with increased—or decreased—distance: first shape, then color and last mass of a body. But the more general value resides in triggering the viewer's projection mechanism through what da Vinci called 'confused shapes.' They suggested new possibilities and stimulated 'the spirit of invention.' The imaginative projection was meant to be key to both the artist's power of invention and the viewer's no less active pleasure of artistic appreciation, whether the projection of recognition or emotion. The standard of mimetic correspondence has lost any rigid grip on reality.

For da Vinci as other Renaissance artists, art and the exploration of the natural world were inseparable, one stimulating the other; his interest in vision was not different from his interest in anatomy and mechanics. The possibilities extended to the artificial world of possible machines. From a cognitive point of view, Gibson and Bateson placed the artistic lesson in an ecological and evolutionary framework, emphasizing the anticipatory guesswork of perception as a form of trial and error (which Popper would turn into the cornerstone of the scientific method). More abstract forms of psychological projection are the ones explored in Gestalt models of higher-order cognition and invention and the ones exploited in the Rorschach test as a diagnostic tool associated with unconscious activity.

Whether or not Leonardo was familiar with Chinese art through Chinese commercial ventures of Italian merchants, the descriptive use of blurred edges already featured prominently in celebrated Song Dynasty paintings and drawings depicting misty and rainy landscapes.

The obverse naturalistic effect is the painter's or picture-maker's use of a sharper area as a simulation of visual focus of attention. The art critic and naturalist John Ruskin noted early how this representational technique was no application of the "truth of nature", or objective standard or truth and realism, but a use of perceptual realism that simulates peripheral vision to guide the viewer's focus of attention.[23]

The assumption is that we are aware of the fuzzy quality of the experiences we associate with those properties of our environment or our cognitive relation to it, e.g., attention and perception of motion. Vagueness of geometric and chromatic intrinsic categorization supports a determinate extrinsic content we conceive as representing sharp properties of the world (exact distance, motion, etc.). Among modern models of perception, recognition of a system may take more holistic and generalized forms that do not require awareness and precise identification of every specific part that might also be an object of perception or representation.[24] Even in

---

[22]Kemp [14].

[23]*Modern Painters*, vol. 1, cited in Gombrich [9], 209.

[24]This aspect relates to holistic aspects of pictorial depiction and violates the phenomenon violates the Fodor-Sober compositionality condition on images.

terms of extrinsic properties or content alone, if we try to deny the intrinsic character of visual fuzziness, accuracy of representation is presented through imprecision. But the standard of perceptual realism assumes a distinction between perceptual recognition of visual fuzziness and, depending on the operative IC–EC link, a not always simultaneous recognition of a determinate content. This is hardly a case of vagueness in blur best understood in terms of precision and overrepresentation.

According to what I have called material realism, imprecision in experience might be accurate of systems that we typically perceive without tonal and geometric definition such as clouds, haze, hair, fur, etc., and not even conceive of them precisely as we do simple solids. This assumption has also become entrenched in the artistic tradition, with fuzziness drawing our attention to the presence of a medium, e.g., J.M.W. Turner and Gerhard Richter. For both what we may call the aesthetics and epistemology of fuzziness relate to the new scientific interests and inventions of their time.

19th-century British landscape painters such as Constable and Turner steeped in old geometry and new experimental science, painting offered a renewed, more realistic observation of the world, freer from academic pictorial schema and pre-conceptions. While Turner also championed dramatic aerial perspective, he worked in the dramatic atmosphere of a new British landscape, with fumes from heated engines and furnaces of the Industrial Revolution, traditional weather conditions and dramatic geological and atmospheric environments the carefully examined in Scotland and Swiss and Italian Alps. His attention turned to the continuum of rock formations and the physical presence of light, fire, speed, smoke and steam.[25] The determinate extrinsic content may be considered, rather than a case of inaccuracy, accurate relative to the extended content that includes perspectival relational properties. Indeed, as instances of interactions with the environment in terms of sensory events and their relevance to our future actions, all perceptions might be judged in terms of accuracy relative to such properties.

For Richter, sharpness or precision of form is an ideological statement, both aesthetic and ontological. It expresses an uncritical commitment to absolute technological reality, unconditioned truth and impartial credibility and objectivity. Like Turner in the 19th century Britain, in postwar East Germany, Richter has offered a sophisticated critique of contemporary representations of Nature and the role of technological means of production. Blur has an equally symbolic and cognitive meaning. Blur is a critical lesson on perception and representation as human practices and constructions, made, partial and conditioned. For Richter blur is, then, a critical thought in painting, a pictorial statement about human standards, representational tools and visible reality.[26]

As matters of representation, aesthetics is a philosophy of mathematics. According to him, there is no exact thinking in painting or in mathematical

---

[25]Hamilton [15].
[26]Richter [16].

calculation, because thinking is identical with carrying out the activities of painting and of mathematical calculation in physics.[27] Production of new forms follows a reaction to prior forms, whether mathematical or painterly. No linguistic formal thinking, in the exact language of 'record-keeping', takes place separate from the performance of the activity. Both are arts that form truths by manufacturing them, side-by-side, supporting or excluding one another. To absolute rightness and truth he opposes human and artificial truth and rightness by construction.[28]

Photography is no different. As a material medium and process, photography is for Richter like exact mathematical physics. It manufactures images with an effect of 'abstraction of its own', but it manages to impose the truth of its likenesses as absolute and unconditioned, its seeing as objective. Photography imposes its images as true and credible, in real space, while displacing the images on painted pictures as artificial, invented, conditioned, inauthentic and incredible, in pictorial space.[29]

Precision of contour and detail becomes the visual mark of photographic realism, of its illusion and persuasion, absolute truth and credence. In photographs, sharpness establishes a rhetorical and regulative standard of realism that is based on a selective differentiation of figures through geometric demarcation. The selective differentiation grants the chosen figures relative importance. Richter wants to expose the photographic imagery as the human construct of an artificial practice informed by the expression of facts and preferences through the application of processes and values. His ulterior aim is to criticize the absolutism of visual truth and rightness that surrounds photographic imagery. And he achieves this by making a 'painted copy' of a photographic image that replaces the invisible exactness of the original with visible blur or fuzziness.[30]

Blur will break the spell of realistic illusion and 'dissolve demarcation and create transition', replacing differentiation with equality within the representation and about it. Blur exposes photography's nature as picture and practice, its transparent objectivity as a visible object, as visible present reality to be documented alongside other visible reality. Every representation is a created analogy.[31] Now the medium is as visible and present as its content, we can both see it and see in it.[32] All the while, content remains clear with (extrinsic) precision and photographic persuasion, with forms fitting more closely and smoothly, 'equally important and equally unimportant.'[33] Only relative to precisely visible reality would painting as representation and analogy be imprecise. For Richter, sharp differentiation is the artificial expression of a cognitive selection, on the basis of conceptual and ontological

---

[27]Richter mentions Einstein, 'Notes, 1962', ibid., 15.

[28]Ibid.

[29]'Notes, 1964-65', ibid., 30–3.

[30]The English edition uses both words for the German expression 'Unschärfe' in the German edition.

[31]Ibid., 55.

[32]Wollheim and Hopkins have stressed his point; see below.

[33]Ibid., 33.

commitments (about preferred categories or properties). To distinguish an object from another is an interpretation based on practical reasons. The flowing transitions in fuzziness avoid such ontological and conceptual commitment. He, again, finds an analogy with physics, with the worldview he associated with Heisenberg's quantum indeterminacy in relation to precise measurement and precise nature of things in quantum mechanics.[34] The reference to quantum physics preserves the ontic-epistemic ambiguity in the status of fuzziness.

In both perceptual and material standards of realism, the assumption is one of categorization of experience of the environment in terms similar to that of the experience of the adequate representation. In particular, pictures are accurate because they are as imprecise as our experience of the same environment. We produce and understand pictures correctly because the ability to recognize imprecision.

Finally, I want to look beyond the artistic context of the cognitive lessons and note very briefly the significance of imprecision with its practical use in specific social contexts. Much image processing is geared towards reducing or eliminating imprecision for the sake of increasing recognition and realism; yet some of it involves the editing task of increasing imprecision. The rejection of high definition might be based on aesthetic grounds aiming at the feel of perceptual realism; technology added to high-definition cameras serve that purpose for instance in porn filming.[35] An important practical purpose in the increase of imprecision is editing for the sake of redacting or hiding visual information from recognition. Typical situations for this type of use involve blurring faces or license plates or any other information that might reveal people's identity for a number of reasons, legal, political or otherwise. A longstanding normed purpose has been censorship (again, often regarding images with sexual content); but it has also been the enforcement legal standards of privacy and protection such as children or witnesses. Examples of this practical kind illustrate a regulative value of imprecision.

---

[34]Ibid., 87. Richter's reference reflects Heisenberg's German cultural prestige free from ideological perspectives. In the original German texts, Richter refers to the precision of photographs typically as 'Genauigkeit' and to blur as 'Unschärfe', and in the interview from 1974 Richter speaks of the way the concept of Unschärfe (translated as 'the term 'fuzziness'') is used in physics with a similar meaning—"In der Physik gibt es, glaube ich, den Begriff der Unschärfe in einem vergleichbaren Zusammenhang" (ibid., 88). The editors' note refers to Heisenberg's well-known quantum indeterminacy principle as 'Heisenbergsche Unschärfenprinzip', the standard German term used at least since 1930 by Schrödinger, along with 'Unschärfenrelation'; originally in 1927 Heisenberg had used the terms 'Genauigkeit' ('determinacy'), 'Ungenauigkeit' ('indeterminacy'), '"schärfere" Bestimmung' ('"sharp" determination') and in an endnote also 'Unsicherheit' ('uncertainty'); it is the last term that is the source of the term 'uncertainty' established in the English expression 'uncertainty principle' by the translation of his 1930 book, *The Physical Principles of Quantum Theory*. The term 'uncertainty' has become also part of the more epistemic vocabulary of fuzzy set theory and logic. In fact, around the same time fuzzy set theory was being applied to modeling quantum indeterminacy, see, for instance, Ali and Doebner [17].

[35]Viewers do not favor the medical look of high-definition displays of skin and its blemishes.

The examples above also give plausibility to the notion and the value of vagueness as practice and in practice. As purposeful activity the objectivity of vagueness is the objectivity of cognitive practices; and with a focus on the particular case of categorization, the practice of vagueness blurs the distinction between the ontic and the cognitive and any associated modes of objectivity. In the last section I extend to discussion of artistic and picture-making practices to the application of mathematics, e.g., fuzzy set theory.

References

1. Huppauf, B. (2009). Toward an aesthetics of fuzzy images. In B. Huppauf & C. Wulf (Eds.), *Dynamics and performativity of imagination: The image between the visible and the invisible* (pp. 235–253). London: Routledge.
2. Hentschel, K. (2001). *Mapping the spectrum.* Oxford: Oxford University Press.
3. Wise, M. N. (Ed.). (1995). *The values of precision.* Princeton, NJ: Princeton University Press.
4. Niepce, J. N. (1839). Notice sur l'héliographie. In L. J. M. Daguerre (Ed.), *Historique et Description des Procédés du Daguerréotype et du Diorama* (pp. 39–46). Paris: Giroux.
5. Cat, J. (2013). *Maxwell, Sutton and the birth of color photography. A binocular study.* New York: Palgrave-Macmillan.
6. Eastlake, E. (1857). 'Photography', *London Quarterly Review*, April 1857, pp. 442–468.
7. Robinson, H. P. (1896). *The elements of a pictorial photograph.* London: Lund.
8. Emerson, H. P. (1889). *Naturalistic photography for students of the art.* London: Sampson Low, Marston, Searle and Rivington.
9. Gombrich, E. H. (1980). Standards of truth: the arrested image and the moving eye. In W. J. T. Mitchell (Ed.), *The language of images, 1974* (pp. 181–218). Chicago: The University of Chicago Press.
10. Gombrich, E. H. (2000). *Art and illusion.* Princeton, NJ: Princeton University Press.
11. Daston, L., & Galison, P. (2007). *Objectivity.* Cambridge, MA: Zone Press.
12. Kemp, M. (1992). *The science of art.* New Haven: Yale University Press.
13. Duhem, P. (1906). *La Théorie Phyique: Son Objet et Sa Structure.* Paris: Vrin.
14. Kemp, M. (2006). *Leonardo da Vinci: The marvellous works of nature and man.* Oxford: Oxford University Press.
15. Hamilton, J. (2001). *Fields of influence.* London: Continuum.
16. Richter, G. (2009). *Writings 1961–2007.* New York: D.A.P.
17. Ali, S. T., & Doebner, H. (1976). On the equivalence of non-relativistic quantum mechanics based upon sharp and fuzzy measurements. *Journal of Mathematical Physics, 17*, 1105–1111.
18. Williams, B. (2002). *Truth and Truthfulness.* Princeton: Princeton University Press.

The examples above also raise plausibility for the notion that the value of vagueness is practical, and its practical. As supposed, it is that by the objectivity of vagueness in the objectivity of certain experiences and which checks on the particular uses of concentration, the practice of vagueness bring this distinction between the data and the cognitive and any associated modes of objectivity. In the last section I voiced my discussion of holistic and entity-making practices to the application of mathematics, e.g., fuzzy set theory.

References

1. Hampton, P. (2001) Toward a statistical theory of concepts. In B. Hampton & C. White (Eds.), *Connectionist perspectives on categorization: The cognitive basis of the brain and the nervous*, pp. 25–60. Hillsdale, NJ: Lawrence R. Hughes.

2. Cohen-Tel, K. (2001) An experimental method. Oxford: Oxford University Press.

3. Wear, M. G. (1997) *Vagueness in nature*. Cambridge, MA: Princeton University Press.

4. Sharp, J. M. (2001) Beauty and imprecision. In L. J. M. Rogers (Ed.), *Concepts of cognition*, pp. 52–80, in *Behaviorism in cognition*. New York: Barnes-Macmillan.

5. Enriques, F. (1957) *The modern theory in mathematics*. April 1957, pp. 143 ff.

6. Rohrmann, H. A. (1969) Prosements. Princeton University Press, London.

7. Knox, R. (Ed.) (1978) *The metaphysics of vagueness*. Cambridge, MA: Harvard University Press.

8. Candlish, E. H. (1988) "Machines of mathematics: Essays in logic and the moving eye." In R. T. McDougall (Ed.), *The History of the language*. pp. 115–130. Chicago: The University of Chicago Press.

9. Gombrich, E. H. (2003) *Art and illusion*. Princeton, NJ: Princeton University Press.

10. Johnston, J. Serviland, P. (2001) *Vagueness*. Cambridge, MA: Zora Press.

11. Kemp, M. (1990) *The science of art*. New Haven, Yale University Press.

12. Dennett, D. (1990) *Darwin's dangerous idea*. New York: Simon and Schuster.

13. Sorensen, M. (2001) *Vagueness: A study of the logic of sorites and vaguenesss*. Oxford: Oxford University Press.

14. Hanfling, J. (2001) *Edge of meaning*. London: Routledge.

15. Raphael, D. (2009) *Wittgenstein*. New York: Oxford University Press.

16. Tye, M. (1990) "On the sorites series of vagueness concepts in the brain." *Journal of Philosophy*, 86, pp. 535–550.

17. Williamson, T. (2002) *Knowledge and its limits*. Princeton: Princeton University Press.

Part 2

Chapter 12
Introduction: Fuzzy-Set Representation and Processing of Fuzzy Images; Non-linguistic Vagueness as Scientific Practice; Scientific Epistemology, Aesthetics, Methodology and Technology of Fuzziness

In Part 2 I conclude the process of tracking objective conditions of vagueness in representation and categorization: from language to pictures, from philosophy to imaging science, from vagueness to approximation and from representation to reasoning. The scientific focus is the application of fuzzy set theory. The main aim of Part 2 is to analyze fuzziness and the application of fuzzy set theory to imaging as a kind of practice applied to the understanding and control of depiction and categorization as other practices. As practices they are enabled and constrained by a variety of resources such as formal and material technologies, habits and skills, interests, standards and limitations.

Fuzziness has been the central formal notion of indeterminacy of linguistic and pictorial representation and the formal notion of its interpretation as either objective categorization or inherent property. Next, I develop ideas from Part 1 and argue that the application of mathematics to the previous conclusions about pictorial vagueness yields lessons about the application of fuzzy set theory, in particular, and mathematics, more generally. We find that fuzzy modeling offers a unified framework and a precedent for treating vagueness in words and images; but the unity of application isn't simple or trivial. Despite a shared interpretive approach based on a formal modeling of categorization, the application of fuzzy sets in the case of images presents distinctive features and challenges. The previous results, I suggest, yield lessons for interpreting fuzzy theory and applying it, especially in pictorial representation and reasoning.

Mathematics is, as a practice, application through and through; even in the practice of the pure mathematics.[1] But, as I have noted, also the application of fuzzy formalism to pictures precedes my introduction in philosophical discussion; it can be found in the analysis and processing of images. The practice of processing images, especially fuzzy ones, is a material extension of formal mathematical practices, offering an empirical extension of its technological objectivity.

---

[1]Cat [1, 2].

© Springer International Publishing AG 2017

J. Cat, *Fuzzy Pictures as Philosophical Problem and Scientific Practice*,
Studies in Fuzziness and Soft Computing 348,
DOI 10.1007/978-3-319-47190-7_12

In image processing the constructed visual representation plays an instrumental role as a blueprint for the control of the image, in particular control of its intrinsic content for the sake of fixing the extrinsic one. After assuming what I have called an IC–EC link, according to which intrinsic features contribute to the recognition of extrinsic ones, the image content, it is easy to consider the reduction of vagueness —even of inaccuracy—in one is taken to serve the purpose of reducing it in the other. Also in the fuzzy case I want to challenge the systematic application of the particular form of the condition I have called the PIC–PEC rule—the imprecision-inaccuracy meta-rule.

If I am right, the discussions above lead to four main conclusions that speak to the specificity of pictures and the practical richness of their mathematical treatments.

(1) In the idealized case of imagery as a well-formed language, the process of formal application is directed towards the possibility of effective control. The aim and practice of control are inseparable from techniques and standards of calculation and representation. That is, fixing pictorial contents is part of a richer practice that involves formal and technological processing. And this practice takes place within contexts informed by aims and standards of categorization, understanding and practical application, which in turn involve rules, hypotheses, judgments, habits, capacities and decisions.[2]

We can identify prescriptive and pragmatic dimensions of the uses of formalism beyond their empirical applications. Here we should distinguish between formal and symbolic aspects of mathematical practices of representation, reasoning and computation. Alongside a system of symbols formalism includes a set of explicit rules for constructing well-formed expressions, some of which function as rules of proof or inference. This is significant because it is the formal dimension that enforces the prescriptive role of mathematical structures, including the ideal of axiomatics.

Part of the prescriptive role of axiomatization is strictly speaking constitutive. Structures and techniques, concepts and performances, act as instruments and expressions of endorsed standards of construction, ex., the determination of a subject matter through the implicit definition of terms, or concepts and the constitution of models and domains. This stipulative function may be considered an instance of the broader pragmatic dimension of formalism.

For example, the pragmatic value of axiomatics in mathematical practice (and as mathematical practice) concern several valuable functions: explicit formulation of theoretical statements and explicit, differentiated, representation of their proof; theory demarcation; theory evaluation; theory communication.[3]

---

[2]On such roles in mathematical application see Cat [1, 2]. In connection with the methodological value of ambiguity, Grosholz has defended the notion of productive ambiguity: A systematic use of ambiguity scientific consists in the joint use of a multiplicity of modes of representation and argumentation in problem-solving and persuasion. See Grosholz [3].

[3]Schlimm [4].

Part and parcel of process of formalizing the cognitive role of vagueness is the application of formal methods to fixing, coordinating and extending content through formal models of categorization. Different formal frameworks are available, and for the sake of modeling and controlling vagueness some involve generalizations of set theory, including, but not only, fuzzy sets. Processes of cognition and technological processing may be formalized in such a way by considering rules and techniques of analysis and synthesis; I argue they are not reducible to mere compositionality. Some such rules and procedures connect what I call intrinsic and extrinsic content. In addition, I argue that they satisfy different criteria of objectivity.

(2) One distinguishing feature of the application of the formalism of fuzzy sets to modeling vague pictorial representation stems from the role of the duality of types of contents in making representation possible and the particular role of categorization in fixing contents. Within the common framework of fuzzy categorization, the application of the linguistic standard is hardly uniform, that is, uniformly direct. Following the analysis in Part 1, I argue that the relation of fuzzy pictorial representation is not generally reducible to a simple semantic relation of partial truth or fuzzy predication at work in the treatment of linguistic cases. The duality of content is captured by two distinctions: a main distinction is between linguistic and pictorial contents, and a more specific one, between intrinsic and extrinsic pictorial contents. This is the issue of relevant standards of truth and objectivity.

(3) Moreover, the distinctions between both kinds of contents—that is, sets of categorizations or features—linguistic and pictorial, intrinsic and extrinsic, raise the question of understanding accuracy of representation as a form of pictorial approximation without approximate truth. Is vagueness a form of approximation? Unlike accuracy of measurement or depiction, vagueness is only a form of approximation in fuzzy set models. The result helps bring out more perspicuously the difference between inaccuracy and vagueness.

(4) Finally, pictorial representation and reasoning are inseparable; in ordinary and scientific cognition, representation—that is, representations and representing —often requires some modality of reasoning or computation and different modalities of visual thinking are common and valuable and require some form of representation. Then, representation without truth suggests a different kind of inference-making rules. If fuzzy representation involves some kind of fuzzy reasoning, it can also support its extended application beyond reasoning for the sake of representation to reasoning from representation, for instance, by means of a fuzzy calculus of pictorial representations. Thinking more broadly also replaces rule-based reasoning.

We need to distinguish between formal thinking about fuzzy pictures and formal pictures about fuzzy thinking. Both involve tools for visual thinking; but diagrams illustrating the symbolic expression of fuzzy set-theoretic concepts and results are not fuzzy pictures with fuzzy design or intrinsic content. Rules must cover the related cognitive tasks that include inference, computation and problem-solving associated with fuzzy visual thinking in perception and pictures. But this is not without implications for the role of truth and thinking. Each presents challenges and limitations when we try to use and understand them. The extension to scientific visual thinking requires distinguishing between fuzzy images and fuzzy pictures. Modeling representation in terms of reasoning or computation is not sufficient for modeling visual thinking based on representation; that is, modeling one the second not a trivial extension of modeling the first.

Scientific processes of cognition may be formalized and controlled through rules and procedures connecting what I call intrinsic and extrinsic content. I argue that fuzzy set theory is a formal practice that formalizability, use and validity of, for instance, IC–EC rules are contextual and restricted accordingly; and that, from (1), (2) and (3), the assumption of their general use and interpretation is misguided.

References

1. Cat, J. (2015). An informal meditation on empiricism and approximation in fuzzy logic and fuzzy set theory: Between subjectivity and normativity. In R. Seising, E. Trillas, & J. Kacprzyk (Eds.), *Fuzzy logic: Towards the future* (pp. 179–234). Berlin: Springer.
2. Cat, J. (2016). The performative construction of natural kinds: Mathematical application as practice. In C. Kendig (Ed.), *Natural kinds and classification in scientific practice* (pp. 87–105). Abingdon: Routledge.
3. Grosholz, E. R. (2007). *Representation and productive ambiguity in mathematics and the sciences*. Oxford: Oxford University Press.
4. Schlimm, D. (2013). Axioms in mathematical practice. *Philosophia Mathematica, 21*, 37–92.

Chapter 13
Application of Mathematics in the Representation of Images: From Geometry to Set Theory

Mathematics has long been applied to images, especially in a number of ways that rely on categorizing visible properties not just geometrically, that is, spatially, but also algebraically or quantitatively. In addition to topological properties of curves and surfaces, metric standards contribute quantitative measures of size and distance. In addition, analytic geometry associates with the sets of points that form geometric curves algebraic equations and functions. The equivalence between both types of representations is established by the fact that they provide modes of coordination, a mapping between two sets of real numbers. To each value of a variable ranging over one set of real values represented by points on one line, or axis, corresponds the value of another on another axis. The curve is a set of points represented by the set of such pairs of numbers, their coordinates. At a higher-level of algebraic structures, we also find group-theoretic properties categorizing and classifying different systems of geometry based on distinctive groups of transformations between sets of points and corresponding invariants.

At the empirical level of applied geometry, surveying and cartographic mapping extend the coordination techniques to surfaces as a set of infinite diagram curves; although typically they only depict the ones corresponding to area boundaries such as coast lines and administrative borders. Coordinate systems in the form of two-dimensional grids are superimposed upon maps and provide a representation, an analysis, of each location (the pair of coordinates are a numerical mapping).

Now, about images more generally: In Part 1 I have discussed how fuzzy set theory has been applied to the empirical (Zadeh) and conceptual (Smith) understanding of linguistic vagueness as a quantitative, conceptually precise model of categorization practices. Visual description and experience take meaningful place only within the context set by a perceptual system. From a categorization standpoint, it makes sense to claim that linguistic and visual indeterminacy and

© Springer International Publishing AG 2017
J. Cat, *Fuzzy Pictures as Philosophical Problem and Scientific Practice*,
Studies in Fuzziness and Soft Computing 348,
DOI 10.1007/978-3-319-47190-7_13

vagueness are inseparable.[1] As a result, the same formal approaches should be, at least heuristically, and have been applied to images, and this in two different ways I call synthetic and analytic.

The *synthetic approach* extends to images the same cognitive focus and the formal resources used to model vague predicates, with the categorization of a particular image by means of its inclusion in a set. Inclusion in the set is determined by a degree of membership, a mapping to a value within the range [0, 1], where 0 and 1 are each assigned to a prototype playing the constitutive and prescriptive role of exemplary instances or standards, a references and guides, as role models. The other assignations are then relative especially to those prototypes.

Fuzzy modeling is *concrete, relational* and *contextual*. It is concrete in its reliance on particular instances. It is relational in that the standards that set the bounds of membership assignations are selected on a comparative, contrastive basis. From a practice standpoint, the two prototypes are adopted for maximum contrast within a variable context that is characterized and limited by material and cognitive conditions, e.g., the availability of potential instances, interests, habits, capacities and background assumptions. Within this context the prescriptive and subjective dimensions of categorization and formalization become objectified.[2] It is also *dynamic*, insofar as the central prototypes may change with the context.

The *analytic approach* considers images as collections of individual points. These perceptual objects or information points are units of visual data to be categorized separately, in terms of a fixed set of perceptual properties.[3] What counts as a unit, e.g., a pixel, is typically established by material and design properties of the medium. On that basis, collective properties of the pixels can be then associated with additional, higher-level categorizations, intrinsic or extrinsic. They are constrained by the aims and categorization schema relevant to a particular context of pattern recognition— examples of problems involving the determination of extrinsic contents include facial recognition and medical diagnosis. Post-analysis synthetic approaches may deploy a number of composition strategies, beginning with the simplest computational assumptions such as set-membership, additivity and linearity.

References

1. Peters, J. F. (2009). Tolerance near sets and image correspondence. *International Journal of Bio-Inspired Computation, 1*(4), 239–445.
2. Peters, J. F., & Pal S. K. (2010). Cantor, fuzzy, near, and rough sets in image analysis. In J. F. Pal & S. K. Peters (Eds.), *Rough Fuzzy Image Analysis. Foundations and Methodologies* (pp. 1–15). Boca Raton, FL: CRC Press.

---

[1]Peters [1], Peters and Pal [2].

[2]I discuss these dimensions in more detail in Cat [3].

[3]This line of conceptual and technological application to the case of images originates in works such as Rosenfeld [4], Pal [5, 6].

3. Cat, J. (2015). An informal meditation on empiricism and approximation in fuzzy logic and fuzzy set theory: between subjectivity and normativity. In R. Seising, E. Trillas, & J. Kacprzyk (Eds.), *Fuzzy logic: Towards the future* (pp. 179–234). Berlin: Springer.
4. Rosenfeld, A. (1979). Fuzzy digital topology. *Information and Control, 40*(1), 76–87.
5. Pal, S. K. (1982). A note on the quantitative measure of image enhancement through fuzziness. *IEEE Transactions of Pattern Analysis of Machine Intelligence, 4*(2), 204–208.
6. Pal, S. K. (1992). Fuzziness, image formation and scene analysis. In R. R. Yager & L. A. Zadeh (Eds.), *An introduction to fuzzy logic applications in intelligent systems* (pp. 147–183). Dordrecht: Kuwler.

Chapter 14
Cognitive Framework of Set-Theoretic Methodology of Analysis and Synthesis: Categorization, Classification and Many Faces of Digital Geometry

Both approaches to so-called visual analysis have set-theoretic expression, based on the assumption that visual cognition takes place at the level of classes, not individuals.[1] The analytic approach has combined classical fuzziness with other generalizations of set theory such as rough and near sets.[2] But a cursory examination of the terms of application of the set-theoretic formalism brings out problems and features that set the pictorial case apart from the linguistic alongside their similarities.

Even the perceived relative homogeneity in individual images, or parts thereof, may be the outcome of a process of post-analysis synthesis: for instance, by applying clustering algorithms to establish classes or else statistical techniques for the calculation of collective features out of the units of analysis. Now, the methodological challenge consists in linking the application of the two approaches, especially with critical and informed attitudes in order to serve best specific aims of representation and reasoning at hand by the techniques and standards at hand.

While set-theoretic approaches may be embedded within other formal frameworks, the latter have become the basis also for classical algebraic alternatives.

Digital geometry is one such basic framework.[3] It is based on the manipulation of numerical algorithms for the sake of representing and reducing blur. The discrete structure of a pixelated image receives arithmetic representation in the form of matrices of values of binary variables. The central formal assumption is the modeling of an image by means of a multi-dimensional system of binary digits, typically two-dimensional arrays ($m \times n$) for grayscale, and three-dimensional ones for color scale.

---

[1] Ibid., and Orlowska [1].
[2] See below and Peters and Pal [2].
[3] See Hansen et al. [3] and Klette and Rosenfeld [4].

© Springer International Publishing AG 2017
J. Cat, *Fuzzy Pictures as Philosophical Problem and Scientific Practice*,
Studies in Fuzziness and Soft Computing 348,
DOI 10.1007/978-3-319-47190-7_14

The algorithms assume a central epistemic distinction, between the perceived or recorded blurred image, B, and the true or desired sharp image, X.[4] Then the form of the problem and the reconstruction task of solving it are to find the digital expression of the true or exact image from the blurred one. X = F[B], where F represents a variety of mappings and techniques. If F represents the so-called deblurring, its inverse represents blurring.

This task is constrained by another epistemic assumption, the conceptual distinction between the exact blurred image, B_{ex}, and the error factor, E, random noise associated with the process of image/data production; one represents the source of imprecision, the other the source of inaccuracy. How is the distinction drawn? It requires both empirical knowledge of the instrument and statistical analysis. We can understand the recorded or perceived blurred image in terms of relations such as the simplest one based on additive linearity, $B = B_{ex} + E$. But then, a fuzzy-set development of this kind of digital analysis and treatment must distinguish between error and uncertainty. This is an epistemic requirement that involves both a conceptual and an empirical challenge.

A perceptual system is a pair formed by a finite non-empty set of sample perceptual objects (standards or exemplars), O, and a non-empty countable set of probe functions, F, representing possible attributes or features of the objects.[5] We can consider perceptual systems particular kinds of information systems, where the universe of objects is a set of data. This combines the cognitive and objective aspects of the system and its relational activity—or interactivity.

We can specify the model of perceptual systems further. The relevant objects are sources of reflected light in the visual field. The may be the image pixels on a digital screen. The mathematical labeling of sensations provided by a so-called visual probe is a requirement on the operative notion of observability.[6] In particular, the role of visual probe functions is to map visual field of points of reflected light sources onto a set of real numbers, assigning feature values; this is the so-called task of feature extraction. In this mapping, signal values are sensations connecting sense stimuli (e.g., in retinal cells) and brain activity (e.g., in visual cortex cells); then probe functions model seeing as transmission processes in perception. They are said to extract perceptual information. Probe functions can map a physical continuum in the environment onto a value distribution over a phenomenological continuum of sensations.

Within the analytic approach and with images modeled as sets of visual data points, the formal modeling of digital technology no longer draws a distinction between image and picture, perception and picturing, anatomy and technology. The approach provides a more abstract and general framework; within it these are all different instances of the same kind of cognitive system. But despite benefits such as unity, abstraction also has costs. I will return to this issue in below. This is part of

---

[4]Hansen et al. [3, Chap. 1].
[5]Peters [5].
[6]Peters and Pal [2].

the larger issue of the unity of framework of conceptual and formal application seeking to accommodate fuzziness in linguistic and pictorial systems.

A few details will show the contrast between the formal framework of practices and concepts of visual categorization and representation and the case of linguistic representation and reasoning. While in the linguistic case some empirical, methodological and subjective dimensions of the categorization process are black-boxed, in the pictorial case, they find formal representation.

From a set-theoretic standpoint, the formal model of cognition at work here extends the application of the linguistic model: a concept or category is introduced extensionally, as a subset of the universe of objects; and to categorize is to assign set membership. The application reaches a higher level of abstraction in extending to cognition beyond perception. A family of concepts (subsets) constitutes abstract knowledge about the universe and the corresponding partitioning enables the classification of objects.[7] A knowledge base is a relational system K = (U, R), where U is a non-empty set of objects and R a family of equivalence relations on U partitioning the universe into concepts or subsets. A subset X is R-definable or R-exact if it's the union of basic categories, which are defined by indiscernibility relations on U. The classification is exact or precise if based on R-exact concepts. If they cannot be defined in the R knowledge base, they are R-rough or vague sets.[8] The exact definition of X in U signifies also the completeness of knowledge about U; then notions of entropy, for instance, provide a general measure of roughness.[9]

A set-theoretic representation provides the operative conception of vagueness with several dimensions of objectivity introduced in Part 1: *formal*, *methodological* and *empirical*. First, the mathematical character of fuzziness—the symbolic relations that help form the relevant concepts—provides a dimension of *formal objectivity*. Second, the application of the formalism is constrained by a more or less broad context of standards, goals, rules, choices, etc. that provide a structure of *methodological objectivity*.[10] Third, as a semantic matter of truth, authors such as Smith locate the objectivity of the linguistic vagueness of a predicate in the properties the predicate represents in contrast with a property of the epistemic state of the language user (that is, in relation to the determinate property conceptualized according to the use of the predicate). Now, when the property of the individual in question, a pixel, is itself a perceptual property, with a physical basis at its source and its detection, the distinction between objective and epistemic conditions collapses. The set structure, through a membership function, models two *empirically objective* states: an act of categorization and the objective cognitive relation it categorizes, namely, a perceptual property between the medium that displays the image and the viewer such as brightness and the tonal contrast that define an edge.

---

[7]Mushrif and Ray [6, 10.2].
[8]Ibid., 10.3.
[9]Sen and Pal [7, 3.6].
[10]For a discussion of the limits of this methodological objectivity see Cat [8].

References

1. Orlowska, E. (1985). Semantics of vague concepts. In G. Dorn & P. Weingartner (Eds.), *Foundations of logic and linguistics* (pp. 465–482). London: Plenum Press.
2. Peters, J. F., & Pal, S. K. (2010). Cantor, fuzzy, near, and rough sets in image analysis. In S.K. Pal, & J.F. Peters (Eds.), *Rough fuzzy image analysis: Foundations and methodologies* (pp. 1–15). Boca Raton, FL: CRC Press.
3. Hansen, P., Nagy, J. G., & O'Leary, D. P. (2006). *Deblurring images. Matrices, specta, and filtering*. Philadelphia: SIAM.
4. Klette, R., & Rosenfeld, A. (2004). *Digital geometry*. San Francisco: Morgan Kaufmann Publishers.
5. Peters, J. F. (2009). Tolerance near sets and image correspondence. *International Journal of Bio-Inspired Computation, 1*(4), 239–445.
6. Mushrif, M. M., & Ray, A. K. (2010). Image segmentation: A rough-set theoretic approach. In S.K. Pal, & J.F. Peters (Eds.), *Rough fuzzy image analysis: Foundations and methodologies* (pp. 10.1–10.15). Boca Raton, FL: CRC Press.
7. Sen, D., & Pal, S. K. (2010). Image thresholding using generalized rough sets. In S.K. Pal, & J. F. Peters (Eds.), *Rough fuzzy image analysis: Foundations and methodologies* (pp. 3.1–3.29). Boca Raton, FL: CRC Press.
8. Cat, J. (2016). The performative construction of natural kinds: Mathematical application as practice. In C. Kendig (Ed.), *Natural kinds and classification in scientific practice* (pp. 87–105). New York: Routledge.

Chapter 15
Conceptual Resources and Philosophical Grounds in Set-Theoretic Models of Vagueness: Fuzzy, Rough and Near Sets

15.1 Set Theory Enters the Picture

The conclusions of my discussion of pictorial precision and certainty in informal philosophical analysis extend to the application of fuzzy formalism. In fact, they do so more broadly. In image analysis fuzzy formalism has been applied as part of different generalizations of set theory: near sets, rough sets and fuzzy sets, and their possible combinations.[1] They are all different kinds of Cantor sets in which membership is established by a law that enables the grouping into a whole. Both fuzzy sets and rough sets generalize classical set theory; and near sets are more general than rough sets.

Also near and rough sets provide general formal concepts for modeling vagueness.[2] In the technological context that has motivated their introduction, vagueness is understood in the more cognitive of the objective senses, in terms of indefiniteness in the practice of classification decisions. Some authors distinguish between two kinds of indefiniteness or ambiguity: feature-ambiguity based on vague definition of region or set boundaries and feature-ambiguity based on indiscernibility.[3] Fuzzy sets provide measures of closeness in membership and truth relative to a feature; rough and near sets provide formal measures of closeness in indiscernibility. Either way, the boundary is established as a matter of decidability of categorization. Thus we can link both sets of notions to measures of similarity relative to a feature or category. Fuzziness is then assessed in terms of degrees of membership based on comparative considerations of similarity.[4]

Nearness and *roughness* are, however, somewhat more general concepts. Nearness is applied to subsets belonging in different sets or clusters; roughness involves exact approximations and extends to objects declared outside the set.

---

[1]Sen and Pal [1].
[2]Ibid.
[3]Ibid.
[4]Cat [2].

© Springer International Publishing AG 2017
J. Cat, *Fuzzy Pictures as Philosophical Problem and Scientific Practice*,
Studies in Fuzziness and Soft Computing 348,
DOI 10.1007/978-3-319-47190-7_15

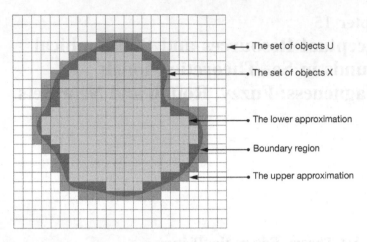

Fig. 15.1 Diagram representing a rough set

Together these notions model notions of indefiniteness, indeterminacy or uncertainty, features that are often considered, mistakenly, equivalent to vagueness or ambiguity. From the formal standpoint, they may be all considered different forms of approximations (see Chap. 20).

Rough sets are characterized in topological terms by the size of the non-empty boundary region of their so-called approximation, or the difference between their upper and lower approximations.[5] Central to those approximations is the notion of partition of the universe into a quotient set of elementary equivalence classes of indiscernible objects with identical feature function/vector values. Based on the basic blocks of indistinguishable or imprecise objects, rough sets are approximated by two exact sets over the partitioned universe corresponding to two exact concepts: these are equivalence classes among their subsets and intersecting sets; one, the lower approximation, is the union of such subsets within; the other, the upper approximation, is the union of intersecting sets with outside. If they are not identical to their lower approximations, the upper boundary provides a non-empty precisely determined boundary region (Fig. 15.1). The boundary region contains a number of sets with an equivalence relation. The measure of roughness at the boundary r = 1-(cardinality of lower approx./cardinality of upper approx.) is a measure of inexactness or incompleteness of knowledge about X.

The notion of rough boundary plays a formal role similar to that of the more general notion of tolerance relation in near sets. In image analysis the objects are the points or pixels into which digital images are analyzed and then placed in the set-theoretic structures of classes defined by indistinguishability relations. The approximations enable segmentations of a query image we want to investigate into non-overlapping homogeneous regions based on specific intrinsic features such as

---

[5]Pawlak [3].

color components or shades of grey of the unit pixels, say, and subspaces of X based on categories of images such as different kinds of systems or phenomena in the relevant classification scheme, for instance, classification of pathologies affecting the same kind of organic structure, in the knowledge archive; the relevant categorizations are based on other, higher-level extrinsic similarities.

The synthetic dimension of image processing rests on the use of these general notions of sets: that is, to the units of analysis within the image and to the classification of image as part of a larger class; images are sets and perception and depiction are class-level, not individual-level, phenomena.

In near sets, disjoint sets resemble each other: description matches correspond to observable similarities. Descriptions of members of near sets are partially matching.[6] Nearness measures the degree of perceived similarity, or, equivalently, of perceived difference between description values, between objects within each single set is higher than across (near) sets. The measure doesn't distinguish between kinds of objects or between objects and pictures. But as a general measure of similarity, nearness obviously includes a measure of co-categorization and, therefore, transparent representation. In fact, it can measure degrees of representation as a relation between objects as members of a set and as a relation between objects as sets.

One method of pattern recognition applies near sets through perceptual tolerance. It identifies a perceptual system and introduces a tolerance relation: disjoint subsets of O X and Y are tolerance near sets if and only if there exist respective subsets of X and Y that resemble each other relative to a subset of probe functions F representing object features and a tolerance class e, the least angular distance so that all points indistinguishable from any x lie with distance e from it. Tolerance is reflexive and symmetric, but not transitive. In this sense, tolerance is a generalization of the relation of indistinguishibility used in rough sets. The distance e may represent, for instance, pixel size as a measure of degree of resolution; each uniform pixel is a tolerance equivalence class. Images can be then analyzed into tolerance classes.

I will be recalling these set-theoretic characterizations of imprecision throughout the rest of Part 2 to address different issues.

Fuzzy sets are near each other if the fuzzy probe function assigning membership grades assign values to at least one object in each within a margin set by the tolerance range, or distance, e. This constant sets the margin of difference in degree of similarity or closeness of description, as in Smith's notion of vagueness, except this is a weaker shared version in respect of categorization, not truth. Nearness is one way to formalize the criterion $V_4(p)$.

My distinctions between intrinsic and extrinsic contents or categorization and between analysis and synthesis apply: Processing is based on the interplay of intrinsic features of units of analysis and extrinsic features of images as units of synthesis. The same framework suggest its relevance to fuzzy processing of visualized data sets (see Part 1), in which the visual design establishes the intrinsic

---

[6]Peters [4].

features or categories and the empirical or theoretical coordinates their extrinsic ones.

The distinction between images as perceptions and pictures is relevant here too. Processing concerns pictures, that is, it involves the formal modeling of images as experienced pictures. If the images are perceptions, processing models are models of cognition and their technological simulation; the units of analysis is a retinal cell or a neuron and functions map interactions between them, as in neural network models with interactive propagation features.

Each approach provides the basis for establishing clustering algorithms that enable the recognition or classification of images.[7] Fuzzy sets enter the synthetic process of image treatment through clustering algorithms such as the fuzzy c-means algorithm. An overlapping partition is typically fuzzy, handled by fuzzy membership functions. In a trivial and broadest sense, a fuzzy boundary is also a fuzzy overlapping partition with its complementary, or union of complementaries.

Another philosophical lesson follows at the formal level: As in the linguistic case, image classification is a holistic, relational affair. Fuzzy membership functions take the form of a matrix, m_{ij}, for n objects and l centroids or potential clusters. Relational, comparative talk of classification replaces monadic talk of categorization. Classification and class definition and partitioning are based, in fuzzy sets, on the contextual and constrained selection of prototypes (centroids). The latter form the subset of the classified objects that help assign weights or feature values, both in the range set by maxima and minima and the centering averages/means. It is because the means may change with the changing sample space that I have pointed to the dynamic, not just contextual, character of fuzzy-set categorization practices.

Successful clustering procedures yield different measures of feature value distances (or degree of inhomogeneity or variation) between clusters (centroids) larger than distances within clusters. Each centroid may help establish a cluster of objects in terms of the minimization of a so-called objective overall multi-cluster function: in terms of a normed distance or difference in feature value weighted by membership values for different centroids. Each object gets membership values relative to each centroid or cluster to degrees constrained by the minimization of the objective function. Classification or categorization involves the application of a unitary fuzzy membership function over the range of centered distances.

The feature value of the centroids is the means of the feature value weighted by membership values. The optimization algorithm consists in iterations of updating recalculations of the initial random selection of the means associated with each centroid feature value in terms of the given membership values. The iterations continue until one converges on an optimum solution with stable values, the c-means, with recalculated membership values showing an increase above a given threshold (a contextual choice).

Rough clusters may be formed out of fuzzy ones. One driving motivation of hybrid rough-fuzzy algorithms is the intuition that the definition of the class is

---

[7]Maji and Pal [5].

vague or uncertain to a degree allegedly best modeled by the upper approximation of a cluster as a rough set.[8]

In rough sets, fuzzy membership becomes relevant in the upper approximation of a cluster: only boundary objects are fuzzified; there fuzziness adds another layer of approximation.[9] Rough sets assign the objects in the lower approximation a membership value of 1; when the difference with the next lower membership reaches above a threshold value, their cluster membership is declared exclusive (a second contextual value, data-dependent, empirically determined by the average highest membership value difference over all objects).[10]

The resulting cluster has a crisp core and a fuzzy boundary around a weighted average. The partition yields a classificatory segmentation, a method of categorizing analysis of image pixels. Distributions of values of some feature must be correlated with their perceived spatial distributions (static or dynamic).

15.2   Philosophical Motivations

I have been tracking and tightening the relation between philosophical and scientific problems, and between language and images. With distinctive mathematical and empirical dimensions, the scientific discussion extends the philosophical. The continuity is not hypothetical and the philosophical examination is also motivated by scientific presentations. Alongside empirical application, the three set-theoretic models of vagueness are motivated by explicit philosophical considerations and goals. In fact, both earlier and more recent discussions include even references to particular philosophical figures.

Cognition in terms of perception, the use of symbolic language, conceptualization and reasoning has been a time-honored preoccupation of philosophical inquiry, and it has been increasingly shared with empirical scientific research. Overlapping projects include scientific philosophy, in which general questions in a certain domain are addressed with empirical or formal scientific resources, and philosophy of science, in which science is the focus of philosophical inquiry. As a point of qualification, scientific philosophy needs not be exhausted by scientific resources, and much philosophy has been inspired but not determined by the systematic and exclusive application of scientific resources; this includes the case of scientific philosophy of science. They gray area between philosophy and science as disciplinary domains is itself fuzzy and projects there feed off resources from both domains of inquiry. The three set-theoretic approaches to what has generally been called vagueness, uncertainty or imprecision have developed in this disciplinary overlap.

---

[8]Maji and Pal [5], 2.2.
[9]Ibid., 2.5.
[10]Ibid., 2.7.

Lotfi Zadeh formulated the basics of fuzzy set theory in the context of comprehensive research on problems of classification, from biology to information patterns.[11] The same problems placed his research in a broader philosophical context that included two overlapping research areas of early 20th-century analytic philosophy, logic and philosophy of language, in particular, multi-valued logic and linguistic vagueness. In those areas Zadeh was inspired notably by a fellow Baku-born, the analytic philosopher Max Black.[12] In a path-breaking essay on vagueness Black echoed Duhem's concern with imprecision at the core of the testing of physical theories and Russell's concern with the application of logical analysis of language to vague terms and offered a quantitative calculus of ordinary language use.[13] Black's formal and empirical approach consisted in tracking observers' linguistic choices statistically and plotting what he called consistency profiles in the ranking of ordered systems. The curves are similar to the curves of degrees of membership in fuzzy sets.

Also the Polish computer scientists Zdzisław Pawlak initiated his work in rough set theory in the context of researching classification; for him it was a general problem in a variety of areas within Artificial Intelligence.[14] As in Zadeh's case, for Pawlak information processing involved a limited role for the applications of classical formal models in the classification of data, or what he called information systems—sets of objects, selected features, data consisting of the features' measurement values and information functions mapping them all.

In the more recent case of near set theory, James F. Peters has characterized his general framework as a combination of views of perception in psychophysics and phenomenology.[15] The view allows the new theory to connect conditions of perception, conditions of similarity and conditions of description in terms of object features.

Peter's account of psychophysics incorporates the basic conceptual apparatus of the original stimulus-sensation model and the revision by the mathematician Henri Poincaré of the original empirical result in the context of his philosophy of mathematics.[16] Peter's conception of a perceptual system is based on the formal concepts of sense inputs (stimuli) and signal values (sensations) out of which the task of feature extraction takes place to identify the object's features. The system is instantiated by measurable states of physical systems (retinal neuron stimulation and visual cortex neuron stimulation), including a process of transmission that is formally represented by a probe function mapping the two sets of values.

---

[11]Zadeh [6].

[12]McNeill and Freiberger [7]; for a more detailed study see Seising [8].

[13]Black [8]. Also Black's references track a combination of scientific and philosophical issues and resources that may be found, in turn, in Duhem, Russell and others.

[14]Pawlak [3].

[15]Peters and Pal [9], 1.5–6.

[16]Poincaré [10].

For Poincaré the phenomenological facts about similarity in the perception of physically continuous systems challenges Fechner's empirical law of psychophysics according to which the degree of sensation is proportional to the logarithm of the degree of stimulus. In particular, the similarity perceived between two sensations fail to satisfy the property of transitivity: sensation A appears indistinguishable from B, B appears indistinguishable from C, yet A appears distinguishable from C. As a result, noted Poincaré, the perceptual continuum cannot underpin arithmetical notions of ordering and continuity; and this shows that the mathematical concept of the continuum, which satisfies transitivity, must be contributed by the mind.[17] The conclusion is part of Poincaré's case against an empiricist philosophy of geometry and his defense of conventionalism. His formal insight motivated Peters' introduction of tolerance relations—in fact, as well as fuzzy measures of similarity.

But Peters doesn't adopt Poincaré's dualist conclusion and, instead, replaces it with insights from Maurice Merleau-Ponty's philosophy of phenomenology.[18] Merleau-Ponty, partly inspired by Gestalt psychology, rejected the dualisms between mind and matter and between thought and language of Cartesian rationalism, the association of knowledge and representation of Kantian idealism, the causal mechanics of behaviorism, and the atomistic picture of sensations of empiricism. Instead, he defended the holistic and embodied character of cognition along with a direct, anti-representationalist account of perception. Peters clearly seeks to make perception central to categorization and to the application of set theory without intellectualist assumptions. He distills the idea that object description and object perception are two sides of the same coin, which he introduces in terms of two axioms of phenomenology[19]:

Axiom 1 An object is perceivable if, and only if the object is describable.

Axiom 2 Formulate object description to achieve object perception.

These axioms, along with the definitions of probe functions and perceptual systems provide the foundation of the near set theory of vision-based cognition and image analysis.

Elsewhere, however, we find the endorsement of the a priori, constructive status of the topological foundation of near and rough sets that recalls Poincaré's position[20]:

'In general, the structure of an object is defined as the set of relationships existing between elements or parts of the object. In order to experimentally determine this structure, we must try, one after the other, each of the possible relations and examine whether or not it is verified. Of course, this image is constructed by such a process will depend on the choice made for the system R of

---

[17]Ibid., ch. 2.

[18]Merleau-Ponty [11].

[19]Peters and Pal [9], 1.5–6.

[20]G. Matheron, in Fashandi and Peters (2010), 4.4.

relationships considered as possible. Hence this choice plays a priori a constructive role (in the Kantian meaning) and determines the relative worth of the concept of structure at which we will arrive.'

The reference to a strict Kantian meaning is, however, inconsistent with Poincaré's and Merleau-Ponty's explicit rejection of idealism. My qualifications in Chap. 1 should make clear that my emphasis on the role of categorization, without a specific account of the activity, is compatible with the constructive and the embodied approaches, as well as the ontic commitment to indeterminate autonomous properties of objects (that is, Merleau-Ponty's and Smith's alternative to the imprecision of semantic relations between language and world).

In the three set-theoretic accounts, then, the different approaches to the philosophical problem of vagueness combine broad philosophical assumptions with formal and empirical scientific elements.[21]

References

1. Sen, D., & Pal, S. K. (2010). Image thresholding using generalized rough sets. In Pal and Peters 3.1–29.
2. Cat, J. (2016). The performative construction of natural kinds: mathematical application as practice. In C. Kendig (Ed.), *Natural kinds and classification in scientific practice* (pp. 87–105). Abingdon and New York: Routledge.
3. Pawlak, Z. (1982). Rough sets. *International Journal for Computing and Information Science, 11*, 341–356.
4. Peters, J. F. (2007). Neat sets General theory about nearness of objects. *Applied Mathematical Sciences, 1*(53), 2609–2629.
5. Maji, P., & Pal, S. K. (2010). *Rough-fuzzy clustering algorithm for segmentation of brain MR images*. In S.K. Pal and J. F. Peters, pp 2.1–21
6. Zadeh, L. A. (1965). Fuzzy sets. *Information and Control, 201*, 240–256.
7. McNeil, D., & Freiberger, P. (1993). *Fuzzy logic*. New York: Touchstone.
8. Seising, R. (2007). *Fuzzification of systems: The genesis of fuzzy set theory and its initial applications*. New York: Springer.
9. Black, M. (1937). Vagueness: an exercise in logical analysis. *Philosophy of Science, 4*(4), 427–455.
10. Peters, J. F., & Pal, S. K. (2010). Cantor, fuzzy, near, and rough sets in image analysis. Pal and Peters, 2010, 1–15.
11. Poincaré, H. (1905). *Science and hypothesis*. London: Walter Scott.
12. Merleau-Ponty, M. (1965). *Phenomenology of perception*. London: Routledge & Kegan Paul.

---

[21]I have examined the case of causality in Cat (2006).

Chapter 16
Analytic and Synthetic Forms of Vague Categorization

Most quantities or measures of interest modeling some form of uncertainty correspond to the intrinsic kind. Fuzzy image analysis is based on the premise that the properties of edge, boundary region or tonal relations in images are not generally represented in sharp terms.[1] In general, the selection of relevant features and their standards is constrained by the aims and standards of the technical practice of analysis or processing and its specific applications, e.g., as formulated by performance indices (see below). They constitute the feature vector that represents the unit of information or perception such as a pixel. One basic feature is tone scale value—represented as a simple scalar or as a vector in basic color space of 2, 3 or 4 dimensions.

The collective form of synthetic representation consists in derived measures of fuzziness over data populations; typically they are first and second-order statistics determining the resolution and precision of images. Examples of first-order statistics features are comparative neighborhood relations of homogeneity and edge value; examples of higher-order ones are Haralick's texture featured based on the spatial distributions of perceptual units measured by joint probability distributions, from angular moments to sum and difference entropies.[2]

The basic, simplest model image for developing the generalized set-theoretic approach is the greyscale digital image. The image is identified with the intrinsic content of a region in the digital medium, the picture, and consists in an arrange of grey values presented by the different individual pixels associated with varying degrees of light intensity on a continuous scale ranging from white, the brightest, to black, the darkest. The different levels form the universe that is represented in a histogram. Then ambiguity or vagueness is introduced as the uncertainty or incompleteness of knowledge about this universe. The stated goal is to provide an objective formal measure of grey ambiguity in the arrays and a technique for reducing it by replacing granulation with segmentation.

---

[1] Pal and King [1, 2].
[2] Maji and Pal [3].

© Springer International Publishing AG 2017
J. Cat, *Fuzzy Pictures as Philosophical Problem and Scientific Practice*,
Studies in Fuzziness and Soft Computing 348,
DOI 10.1007/978-3-319-47190-7_16

The values on the in-principle continuous scale enable a gray value gradation over an array of contiguous pixels. This is the property within or between dark regions that mid-19th-century photographers valued as the half-tones on which much richness of detail and realism depended. But vagueness should not be confused with the gradation itself, which mathematically can be modeled otherwise, for instance, by gradients. The phenomenology of distributed values above cannot identify the representational property of blur.

In Part 1 I have suggested the application of fuzzy versions of formal treatments in models of dynamic vision models. Each treatment provides models with empirical and technological strengths and limitations. Like treatments of blur, they typically come in two related formats, analytic and statistical or probabilistic. An example of analytic treatments is the geometric approach based on the so-called Gabor wavelets, special Gaussian treatments of spatial frequencies; the Gaussian distribution, while providing a solutions to wave equations also provides a statistical basis. The other statistical or probabilistic approaches are Principal Component Analysis (PCA) (and variants), Linear Discriminant Analysis (LDA), Gaussian mixture estimations (especially, expectation maximizations), Kalman filters, Bayesian belief networks and hidden Markov models.[3]

Instead of the features modeled by analytic and probabilistic models, the kind of tonal vagueness known as greyness ambiguity is a feature (category) of a boundary. That is, it's a feature of yet another, topological, feature that is in turn identified in a topological region with a finite area and geometric shape—both presumably uncertain too. Still, the concepts of pure geometry only help categorize our experiences. In the phenomenological context of images, we deal with empirical geometry, the sensible categorization of experience. The boundary line that distinguishes two visible regions has an equally visible empirical identity in the form of a sharp contrast between homogeneous tone intensities, such as black and white, dark and bright. But the grey value gradation prevents the appearance of a sharp enough contrast, a boundary location. This is the so-called greyness ambiguity of the boundary that other visual cues lead the viewer to expect but cannot locate.

As a matter of objective possibility, not just incompleteness or uncertainty, the indefinite boundary can lie anywhere in the gradation.[4] Recall that to measure the ambiguity, set theory offers two formal criteria: (1) vagueness due to fuzziness of the boundary (the feature associated with the application of a category for which an instance is available) and (2) rough resemblance due to indiscernibility between gray levels responsible for granulation. The construction of specific rough-set criteria such as information entropy measures is an example of a goal-oriented practice; it is constrained by desiderata on formal representation such as non-negativity, continuity, sharpness, maximality, resolution, symmetry, monotonicity and concavity.[5] Some are formal constraints, others are formal but also

---

[3]See their application for instance in Gong et al. [4].

[4]Sen and Pal [5, 3.1].

[5]Ibid., 3.9–10.

motivated empirically, for instance, in terms of feature structure and maximum and minimum conditions.

As a synthetic cognitive phenomenon, indefiniteness, vagueness or undecidability, to use terms authors take to be equivalent, can be traced to the outcome of analysis into the vagueness of greyness levels in individual pixels, even when the ambiguity is due to the indiscernibility between contiguous grey levels. The analysis applies then, more synthetically, to the aggregated collective account of greyness levels represented in a histogram, where the regions are identified and aimed to be distinguished further—as determined by contextual cognitive and practical standards and interests.

Rough sets, in addition to fuzzy ones, help represent the homogeneous element identified in the phenomenology of blur in unaided perception and in its technological extension. The interpretation of blur in terms of overrepresentation can be then understood in terms of either kind of overcategorization, empirical or theoretical—relative to assumptions or background knowledge.

A number of related segmentation methods such as clustering and multilevel thresholding have been introduced in order to control grey ambiguity; they are in effect techniques for minimizing the adopted measure and differentiating grey levels into separate regions with more homogeneous contrasting levels, based on pre-set target threshold values. The simplifying aim is to increase precision of features such as boundaries by increasing simultaneously homogeneity and disjointness in sets of values. This kind of operation is known as feature-extraction. On the greylevel histogram, segmented regions are expected to correspond to regions in the vague image. Extraction then relies on considerations of resemblance. For instance, as in the case of blur, theoretical or perceptual habits lead to an expectation of a recognizably sharp visual boundary. In the visual case, such precisification techniques are based on pictorial desiderata in the intrinsic content of the image are often based on background knowledge and cognitive constraints that help assign the features of extrinsic content through IC–EC rules.

References

1. Pal, S. K., & King, R. A. (1980). Image enhancement with fuzzy set. *Electronics Letters, 16* (10), 376–378.
2. Pal, S. K., & King, R. A. (1981). Image enhancement using smoothing with fuzzy set. *IEEE Transactions of Systems Man and Cybernetics, 11*(7), 495–501.
3. Maji, P., & Pal, S. K. (2010). Rough-fuzzy clustering algorithm for segmentation of brain MR images. In S. K. Pal & J. F. Peters (Eds.) 2.1–21.
4. Gong, S., McKenna, S. J., & Psarrou, A. (2000). *Dynamic vision.* London: Imperial College Press.
5. Sen, D., & Pal, S. K. (2010). Image thresholding using generalized rough sets. In Pal and Peters (Eds.) 3.1–29.

Chapter 17
From Intrinsic to Extrinsic Vague Categorization and Content

Above I have focused on the treatment of intrinsic image features. In the process, I have also noted that aspects of their categorization such as ambiguity and of their control such as threshold values are associated with extrinsic features. In fact, the representational function of pictures is determined by the accuracy of their extrinsic content. This is the dual problem of *content development* and *content coordination*.

Content development builds originally on intrinsic content, the recognized features in intrinsic categorization of the picture; but to do so requires a coordination rule that links extrinsic to intrinsic content, an IC–EC rule. To the extent that the IC–EC rule provides a mode of inference, we can claim that it enhances (reasoning- based) representation through (representation-based) reasoning. Notice that the validity of the inference is based on the validity of the application of the rule, which in turn is based on conceptual/empirical assumptions of limited contextual relevance. Here I don't distinguish between reasoning, inference and computation. The relevance of specific theoretical, empirical or conventional interpretations of the IC–EC rule depends on the context of production of the image. Out of IC–EC links one can apply further links of the form EC–EC′, and so on.

For example, for medical diagnostic purposes, one type of rule is an explanatory hypothesis with empirical support positing a causal link between certain types of breast cancer and sudden increases and asymmetries of local temperature distributions in thermograms. We see tonal distributions, but with the association encoded in the rule the intrinsic information is categorized further in terms of the recognition of cancer.

Another type of IC–EC rule for addressing content development takes the form of a compositional strategy. When only external content of the parts is available, the constructive strategy of the extrinsic content of the whole involves an IC–EC rule— or an EC–EC′ one—that depends on the coordination of subunits of visual information through their individual partial IC–EC rules. These rules satisfy the following compositionality condition: for each unit i, $(IC - EC)_i = (IC_i - EC_i)$. They are applied along with auxiliary coordinating techniques that involve stitching rules and constraints based, for instance, on features of the medium of representation and a code of conventions.

© Springer International Publishing AG 2017
J. Cat, *Fuzzy Pictures as Philosophical Problem and Scientific Practice*,
Studies in Fuzziness and Soft Computing 348,
DOI 10.1007/978-3-319-47190-7_17

Galileo's drawings of the moon already followed this strategy, since the field of view of his telescope couldn't accommodate the entire image of the moon's bright side; a more modern example is the recent mapping the dark side of the moon. In the medical sciences, the computational source of brain imaging, especially functional-MRI imaging, is currently a popular instance of this synthetic process. At the very basic level of MRI imaging, however, the reliance on statistical calculations incorporating information from populations of similar individuals and statistical criteria leads to a controversial relation between image production and image interpretation and trust.[1]

Conversely, solving a problem of content coordination may contribute to solving the problem of establishing a picture's intrinsic content. In a given context of picture-making or interpreting, the identification of the extrinsic content and an IC–EC link help determine by selection the recognized and processing of properties that make up the relevant intrinsic content. A synthetic strategy, then, may precede and inform the task of analysis. Compositional strategies may involve the role of such top-down constraints.

Examples of the strategy go back to the development of 15th-century mechanical, step-by-step, rule-based drawing techniques understood as copying or re-presenting: the orthogonal grid structure in the arrangement of glass panes in windows became adopted as a tool for picture-making by means of mapping by geometrical analysis into smaller areas. Then the perception of the whole determines the representational content of the parts (top-down analysis), while the ensuing representation of the parts will, in turn, yield an adequate representation of the whole (bottom-up synthesis).

The standard of adequacy of representation at work is the kind of perceptual realism set by the experience of looking out a window, which the drawing aims to reproduce through the geometry of grid (in lieu of window) and the rules of classical perspective. This procedure is an instrumental, constructive reversal of the Sober–Fodor image criterion. Modern pixelation is the digital limiting version of the classical grid-based representational technique.

From a methodological standpoint, content coordination serves the epistemic function of evidence. This is analogous to the case of hypotheses supported by empirical data provided one can assume an auxiliary hypothesis linking the relevant information or concepts in the data, e.g., running a fever when the thermometer detects a temperature of at least 100.4F.[2] As I discussed in the note on data, above, the evidentiary value of data is a complex and contextual matter that can hardly be reduced to a general formal rule.[3] The rule is always embedded in an opaque context of practical and cognitive constraints.

---

[1]Bogen [1], Machéry [2].

[2]The role of this kind of auxiliary hypotheses and their status as a vehicle for values and biases is discussed extensively in Longino [3].

[3]Leonelli [4].

The relevant result is that content development is a form of image interpretation by categorization; and thus it may present some form of fuzziness. Image processing in practice, however, is typically based on the auxiliary assumption that content development is based on content control through the minimization of vagueness. The alternative I have mentioned includes coordination of categories based on the positive role of fuzziness, for instance in the increase in blurring for cognitive, aesthetic and practical purposes—such as legal protection of information. Criticism from early photographic realists and the methodology for image analysis suggests that the formulation of the sub-rules is general or that their application is systematic.

Behind blur reduction lies a specific form of IC–EC rule linking the precision status, P or I, of intrinsic and extrinsic contents, namely, sub-rules PIC–PEC and IIC–IEC: they assume that the precision status of each type of content is linked to that of the other in a relation of identity. Only definite or precise intrinsic content guarantees precise extrinsic content; at the very least, we may assume that intrinsic precision sets an upper limit to the possible vagueness in extrinsic content—since intrinsic vagueness only yields a measure of additional extrinsic vagueness.

For all these reasons the PIC–PEC rule, when in fact eliminating uncertainty in the treatment of the image, may be wrongheaded as a general strategy as it may be cognitively and representationally counterproductive. From the point of view of possible rules, the degree of precision of one level of content does not necessarily or sufficiently determine that of another. Their dependence may go either way and is bound by context. Here are some scenarios:

(a) The most common case is the basis for the need of specific, but not universally applicable, IC–EC links. In the absence of additional constraints, precise intrinsic content—what the viewer sees—cannot fix the categorizations in the extrinsic content—what he sees in it. It cannot fix ambiguity, before the additional challenge arises from underdetermination of a set of possible precise extrinsic categorizations —whether the dot on the screen "is" a star or a tumor, a certain person, etc. Even if we leave aside indexicality in tracking content—and challenges concerning the reliable determination of reference—, the general form of the schema for IC–EC rules enabling pattern recognition must accommodate the fuzziness in extrinsic categorization. Each context of image production and interpretation might suggest different specific instantiations with empirical or heuristic value.

(b) In a common type of case, blur either in perception or depiction is identified in relation to the standard or expectation of a precise property, such as a tonal or geometrical arrangement. That expectation may be suggested by habit, also by a contextual interpretive assumption about a corresponding sharp property in an object of depiction (extrinsic content). In one case we are dealing with precision in perceptual representation, in the other with precision of "theoretical" categorization. Precise extrinsic content can be then suggest a top-down condition on the precision of categorization that establishes intrinsic content.

(c) Another type of case is the one Victorian perceptual realists and subsequent cognitive psychologists have established: that a variety of extrinsic perceptual categorizations rely on an intrinsic perceptual cue involving fuzziness. Determinations of distance, focus of attention, texture, volume, even the recognition significant structural detail depend on the absence of sharp contrast and the homogeneous segmentation supporting it. It might not, of course, be sufficient, and, as cases of type (a) suggest, a context for the constraints is required. Nevertheless, it suggests a cautionary note against identifying the sharp/fuzzy dichotomy with the signal/noise one. From a cognitive standpoint, information-based entropy measures are no reliable guide to salient content.

The role of context suggests the incompleteness and unreliability of any rule-based automated procedure. The three scenarios just outlined involve a motivated choice of what counts as the effective subset of the knowledge base, the range of effective conceived possibilities and connections. Constraints that make possible different degrees of indeterminate representation, additional uses of reasoning and the evaluation of outcomes depend on additional inputs from local constraints and external operation. Self-setting automated systems cannot anticipate all the relevant constraints. The contextual practice of formal and machine categorization is typically supplemented by an extended external context of interpretation and use. Fuzziness, then, is deeply contextual, with multiple meanings and serving multiple purposes in different kinds of activities.

References

1. Bogen, J. (2002). Epistemological custard pies from functional brain imaging. *Philosophy of Science, 69*(3), S59–S71.
2. Machéry, E. (2014). Significance Testing in Neuroumagery. In J. Kallestrup & M. Sprevak (Eds.), *New waves in the philosophy of mind* (pp. 262–277). New York: Palgrave Macmillan.
3. Longino, H. (1998). *Science as social knowledge: Values and objectivity in scientific inquiry*. Princeton, NJ: Princeton University Press.
4. Leonelli, S. (2013). Classificatory theory in biology. *Biological Theory, 7*, 338–345.

Chapter 18
Pictorial Representation and Simplicity of Categorization

The premise and focus of this book is the basic role of categorization as a practice in linguistic and pictorial representation—also acknowledging other roles derived from or alternative to representation. I do not take a stand on the explanatory role of categorization as a fundamental cognitive activity or its associated specific mechanisms, capacities and behaviors, nor on its semantic and ontic significance in relation to the "world." I have explored its role as the precondition for the practice of application of formal fuzzy set theory, and have defended it as a general objective model able to accommodate the empirical objectivity of instantiated properties and the epistemic objectivity of cognitive states and operations.

In the linguistic case, on standard views, predicates manage to symbolize a simple, direct and specific judgment of categorization; and they do so regardless of their degree of definiteness. As a consequence, they warrant the truth value of a proposition with predication. This is the basis for the application of formal fuzzy sets to model linguistic vagueness in terms of degrees of truth. Recall from my discussion in Chaps. 3 and 4 that linguistic symbols, with their serial structure are just symbolic denotative devices whose content, unlike that of pictures, is visually opaque, that is, independent, except by convention, of their visual, spatial structure and other visual features.

In the pictorial case conditions become more complicated in at least three sorts of cases. Symbolic use cuts across the role of pictures and language: analogical pictures and diagrams, include elements their contribute their content symbolically and the interpretation of pictures often depends on the auxiliary role of linguistic symbols. To articulate how the pictorial case challenges the unity of application of fuzziness criteria, including fuzzy set theory, I have appealed to three distinctions: *image/picture, intrinsic/extrinsic* and *truth/fit*.

Unlike the linguistic case, in visual (and auditory) fuzziness, the distinction between perception and depiction, image and picture, is relevant to understanding the variety of conditions and practices in which different uses and significance appear. They include the role of a number of constraints, including the more normative and pragmatic role of goals and standards. Key to such practices of understanding, use and control, including in formal and technological treatments, is a distinction between

© Springer International Publishing AG 2017
J. Cat, *Fuzzy Pictures as Philosophical Problem and Scientific Practice*,
Studies in Fuzziness and Soft Computing 348,
DOI 10.1007/978-3-319-47190-7_18

what I call intrinsic and extrinsic contents. The role of intrinsic categorization is additional and conducive to the to extrinsic level; and in the account I have sketched out, even the latter is involved in sort of dual instantiation or, to put it in cognitive terms, co-categorization, of the picture and of the external target; the latter provides the extended, external element of extrinsic content.

Pictorial representation is a function carried out by means of an IC–EC relation in an opaque context of different sorts of conditions. It is the IC, not the particular picture bearing it, that does the representing (of course in practice we may say that people, not pictures, represent or whatever is functionally equivalent). Therefore, the relation of representation isn't simply reducible to simple categorization and, in turn, the categorized picture doesn't stand to its target content in a semantic relation of predication and truth. The status of pictorial representation and the relation of representation depend on the image being an object of categorization and predication. And none of those relations constitute truth, not do they need to be truth bearers to perform cognitive functions, only a weaker form of veridicality, accuracy, etc.[1]

Because pictures cannot be systematically transcribed into a thousand words in a system with linguistic semantics and syntax, much less into true or false statements about the picture's external targets, the linguistic model doesn't help.[2] It has helped only to highlight the role of categories and acts of categorization. The fit or accuracy of the picture depends, nevertheless, on the properties or categories associated with it and its contents, and the problem of relevant units of representation, composition and holism mentioned in earlier chapters is vexing.

Depiction depends often for accuracy of fit on the auxiliary role of linguistic. The fact that symbolic elements often mediate the content of pictures cuts both ways; on the one hand, it complicates the understanding of how the two kinds of relations or practices cooperate in determining content; but on the other, it suggests a broader kind of relation. To claim they both function like demonstratives will not work for pictures, or for propositions. The distinction between images in perception and pictures matters again (even if pictures depict on perceptual grounds). Neither statements nor pictures might be fully involved in a semantic relation to denoted targets without cognitive conditions.

To be clearer, the focus on categorization reveals a distinctive degree of complication in pictorial representation that is only compounded by the distinction between IC and EC. But the IC/EC distinction has contextual grounds and meaning, in particular at the service of representation. I introduce it on the basis of a cognitive distinction between two levels of perceptual recognition (categorization): one identifies visible properties of the picture that are seen, the other tracks properties that are seen-in or seen-as. Distinctions such as object/background and associated features such as object-background border imply some form of IC/EC dichotomy.

The simplest form of the distinction is in terms of an intentional relation EC is declared to bear to the external target subject of representation, whether a type or its

---

[1]Goodwin [1].
[2]Perini [2].

token (or through it). Specifically, EC consists of the subset of categories or properties making up the picture's IC that is identified by that relation; and the relation is one of perceived similarity or co-categorization (and exemplification by the token picture and target).

I admit that the IC/EC dichotomy may be challenged on a number of conceptual grounds. For instance, in debates over representationalism about the nature of perceptual images, a distinction like IC/EC is understood as a questionable form of opaqueness. The viewer would be aware of the image itself, a separate phenomenological experience, rather than of its intentional content. Instead, proponents of representationalism, or intentionalism, defend that images are transparent, that IC is EC, all intrinsic properties are properties of the object in the visual environment. I have argued above against this sort of approach to blur as overrepresentation or inaccuracy rather than vagueness.[3]

Another denial of the IC/EC dichotomy has the form of representationalism about perception and depiction on the grounds that the difference is a difference between kinds of representations and EC is a contextual selection of IC properties on grounds of the contextual interest of the external object or the criterion of representation that singles it out; while the rest of IC properties constitute what I call the incidental, or surplus, EC including a variety of possible entities or properties, concrete or abstract.

But the weaker form of representationalism is compatible with my contextual distinction. Whatever constraints establish the intended or interpreted EC in the mind of the picture-maker or the viewer in terms of a subset of visible IC categorizations, EC provides the relevant kind of content associated with the external system. I call the extrinsic content external whenever EC is instantiated by the relevant system in the context at hand and not just by any set of standard tokens associated with each particular categorization in EC that helps recognize the property in the relevant system at hand or, equivalently, categorize it.

Vague categorization and representation are only particular cases and conditions of a larger class of relations that include similarity, reference, classification, etc. This feature characterizes fuzzy depiction. It is not directly equivalent to an elementary case of fuzzy predication, degrees of membership and linguistic truth. For a representation relation involving a picture and its object, vagueness of categorization and truth would have to accommodate the possibility of a relation between two partial-truth conditions associated with the two levels of fuzziness, intrinsic and extrinsic (one about the picture, the other about the intended object); but neither condition is in itself reducible to a single more basic semantic relation of truth. Nor is the composite that distinguishes pictures from images in perception. Semantics of truth and representation are not identical or entirely equivalent relative to their respective roles and criteria.

This is a potential problem for the standard interpretation of fuzzy set theory when applied to pictures, since it relies on semantic conditions for fuzzy predicates in terms of degrees of truth. I see two alternatives. One is to try to subsume depiction under this semantic relation. The other is to let go of the semantic requirement, at least in the standard ontic version.

---

[3]Siegel [3], Allen [4].

In the first option, we can try to construct a propositional truth condition associated with pictorial representation, but only derivatively; it involves the truth value of a statement about the relation of representation—and shared categorization—, which we may call a transcription or representation report.

In the linguistic fuzzy model, if S is M to degree n, $M(S) = n$, where n lies in [0, 1], then the proposition $p = M(S)$, that S is M, is true to degree n.

In other words, if T is the semantic function that maps p to a range of degrees of truth n (partial truth values) [0, 1], $T(p) = T(M(S)) = n$.

Now, if instead of the truth of propositions we turn to pictorial representation, PR, as a property of a picture P, then $PR(P) = PR(P, Q)$ a function of a relation between P and a target external content Q, e.g., someone's face. The relation may be given a set-theoretic expression as a relation of feature indistinguishability or similarity. On that basis, IC and Q may be characterized in terms of set similarity. P instantiates the geometric and chromatic properties M that make up its IC and Q instantiates $M' = EC(P)$, P's extrinsic content, a face, based on some IC–EC link that associates the visible properties M we see in P with the properties M' of a face we see (recognize) in P as an added level of categorization. The extension of visible content to extrinsic relational properties or categorizations takes place via IC–EC links in the form of conventions, theoretical hypotheses or analogies, e.g., intrinsic features may be seen reconfigured into a face-like arrangement.

P, we often claim, is represented by the categorization of its properties, M and M', PR (M, P) and PR(M', P); and, vice versa, it can represent properties, and we can say it represents M and M', PR(P, M) and PR(P, M'). Here the categorization approach exhibits the potential symmetry in the relation of representation as well as in any relation of visible resemblance that might play a role.[4]

The asymmetry lies in the contextual choice and use of whose visible properties perform the function of representation. It also lies in the judgment that the particular picture represents another particular system, or that the picture as a token of a type represents a token of another type and derivatively it is intended or understood to represent the type whose token exemplifies or refers to (thinking of reference or denotation by the particular token bridges over the gap between the representational role of images and linguistic predicates).

The representation of types, kinds or classes is prominent in the sciences in accordance with the aim to represent populations, properties or generalizations involved in phenomena and their explanations. As I noted above, P doesn't instantiate M' in the same way Q instantiates it, except indirectly and relationally, as contextual subset of relations between intrinsic properties and others associated with the categorization of other actual or possible entities and a standard instance. While we can state M(P), M'(P) has a different kind of meaning as subject of a different sort of categorization; P exemplifies an extended, external type of categorization as a type of picture of that face-like property M' of Q if M and M' are close enough, that is, if P and Q are similar enough in terms of M or M'.

---

[4]Goodman [5], Gentner et al. [6].

To accommodate this duality of recognized content in perception I have adopted a broad approach to objective categorization, according to which categorization describes instantiated properties and acts of categorization as modes of representation by classification. Some forms of categorization place the properties of the picture in a set of similar pictures or prior experiences. The new category refers to this kind of selective reconfiguration of properties otherwise instantiated. Instantiation, in this sense, is similar to what Goodman called exemplification.[5] This is, for instance, how fictional and theoretical entities are portrayed or visualized.

So, if $M = IC(P)$ and P is M to degree n, $M(P) = n$, $M'(P) = n'$,
$PR(P, M) = n$ and $PR(P, M') = m'$
Q is M to degree n', $M(Q) = n'$
and P represents P to degree m, $PR(P, Q) = m$,
then
$PR(P, Q) = F(M(P), M(Q)) = F(n, n') = m.$[6]

Alternatively, if $M'(P) = l$ and $M'(Q) = l'$,
$PR(P, Q) = G(M'(P), M'(Q)) = G(l,l') = k$,
with, one may assume, $m' = m$ (even if $F \neq G$, $l \neq n$ and $l' \neq n'$).
Still, T(P) is not defined. Instead, we have T[PR(P)] for the report $p = PR(P)$, where $T[PR(P)] = T[PR(P, Q)] = T[F(n,n')] = m$.

This conclusion matters first as a point of distinction between linguistic and pictorial fuzziness as a model of vague representation. It is compatible with the alternative of weakening the semantic condition on fuzzy predicates as part of vague propositions. The semantic standard provided by the link between determinate predication and truth would operate, for instance, as a semantic prototype for an extreme instance of the more abstract, weaker notion of correspondence or fit. A more radical alternative consists in a weaker foundation for fuzzy set theory based on categorization without truth, without a commitment to modeling formally our ordinary linguistic practices, or else to reconceptualizing truth accordingly. The substitute central standard could be reliable categorization, with a specific standard of reliability.

Second, the conclusion matters to distinguishing images from pictures, which fuzzy set theory fails to do on its current foundation and its emphasis on modeling and simulating fuzzy perception, specifically on pattern recognition.

Third, the conclusion matters to understanding what approximation can be for pictures, even as a form of accuracy, especially without reducing vagueness to accuracy of truth.

---

[5]Goodman [5].
[6]Question: What's the relation between n, n' and $m' = PR(P, M')$?

Fourth, the conclusion matters to the semantic grounding of rules of inference based on pictorial representations and their construction for an extended and diversified model of pictorial reasoning.

I address the last three points in the final chapters.

References

1. Goodwin, W. (2009). Visual representation in science. *Philosophy of Science, 76*, 372–390.
2. Perini, L. (2005). The truth in pictures. *Philosophy of Science, 72*, 262–285.
3. Siegel, S. (2010). *The contents of visual experience*. New York: Oxford University Press.
4. Allen, K. (2013). Blur. *Philosophical Studies, 162*, 257–273.
5. Goodman, N. (1976). *Languages of art*. Indianapolis: Hackett.
6. Gentner, D., Holyoak, K. J., & Kokinov, B. N. (Eds.). (2001). *The analogical mind. Perspectives from cognitive science*. Cambridge, MA: Bradford Books.

Chapter 19
Fuzzy Visual Thinking: Interpreting and Thinking with Fuzzy Pictures and Fuzzy Data

In this chapter I pick up on previous results and extend the discussion in Chap. 5 to tackle these questions: How do pictures and their fuzziness enter these activities? How can fuzzy set theory accommodate fuzzy visual thinking?

Set theory can be used in the formalization of reasoning from images or with images associated with the representational and inferential use of symbolic diagrams, also in the formalization of thinking in a broader set of tasks that include computation and problem-solving using both fuzzy diagrams and fuzzy analogical pictures.[1] Whether in perception generally or in the use of pictures, performing cognitive tasks involves a variety images, some of them are fuzzy perceptions, others are fuzzy pictures. As I do in Chap. 5 in my broad discussion of representation and thinking, also here I want to include visual thinking in scientific contexts. I should note also that in many situations involving no scientific purposes or projects we think with scientific technology and information such as geometric graphs, photographs and quantitative data. My suggestion is that fuzzy set theory can contribute its formal and empirical resources not just to furthering science and technology, but to understanding them too. But such understanding is not fixed by the application of formalism alone; instead, it relies on interpretation that calls for critical examination. Exploring the limits of empirical representation and reasoning in quantification and valid inferences is only one project, e.g., so-called evidence theory within the generalization of probability theory known as possibility theory.[2]

We can safely acknowledge that thinking with different kinds of fuzzy pictures in different situations amounts to engaging in different kinds of practices, with diverse interests, norms, skills, media and technologies. As a consequence, again, an adequate understanding the uses of pictures cannot be reduced to a single acontextual model, in particular, to understanding perception.

---

[1]On the role of diagrams see, for instance, Larkin and Simon [1], Perini [2], Shin [4], Goodwin [5], Blackwell [6].
[2]For a brief introduction see Klir and Yuan [7, Chaps. 7 and 9].

© Springer International Publishing AG 2017
J. Cat, *Fuzzy Pictures as Philosophical Problem and Scientific Practice*,
Studies in Fuzziness and Soft Computing 348,
DOI 10.1007/978-3-319-47190-7_19

The emphasis on rules of reasoning discards the radical semantic alternative to the theory's foundation that I have suggested in this chapter. It also suggests separating pictorial thinking more generally from pictorial reasoning, where reasoning is often taken to consist in the application of rules of inference. Yet, pictorial reasoning depends on symbolic and diagrammatic elements and on more or less restrictive semantic conditions for supporting valid inferences. The role of rules blurs the difference between reasoning and computation, especially when it relies on inference and proof. But there's no gain here in defending the distinction; the heuristic benefit comes from turning to the more inclusive notion of thinking.

19.1 Visualized Thinking

It's then worth distinguishing between five related types of cases: (1) visualizations of fuzzy-set mathematical structures and thinking with them, (2) visualizations of fuzzy reasoning, (3) fuzzy-reasoning-based perception, (4) fuzzy-set models of fuzzy visual thinking and (5) fuzzy-set models of fuzzy-visual-thinking-based perception. Here I want to draw attention to the third type and point to the foundational challenges and research opportunities that are suggested by the results in previous chapters.

First let me point out how pictures play a role in case types (1) and (2).

The most basic concept in fuzzy set theory is the degree of membership of an object in a subset. It is an algebraic relation mapping objects onto real numbers. In addition to the standard symbolic notation for numbers, variables, functions, operations on them and relations between them, the mapping has prompted two related pictorial representations stemming out of the curve maps out in space the membership function. To speak of geometry can be misleading. From a pictorial standpoint, it is the graphic expression of the curve, the concrete spatially-extended and visible figure that matters, not the mathematical structure.

The first application of the membership diagram is in the construction of extended Venn diagrams (Fig. 19.1).[3] Classical Venn diagrams for classical set theory are effectively closed curves representing the common value 1, full membership, of the membership degree of its enclosed members. Perhaps it is more intuitive to picture the same feature as the rectangular area enclosed by the straight line representing the membership function for objects with full membership— actually, a step function. In the extended fuzzy version, the straight line assigning value 1 is replaced by variable curves. As in the classical diagrams, we can use the extended ones to visualize the corresponding set-theoretic operators.

The second visualization corresponds to the two and three-dimensional versions of the n-dimensional hypercube in which each n-dimensional point represents a set

---

[3]The diagrammatic structure was introduced by Zadeh in Zadeh [8].

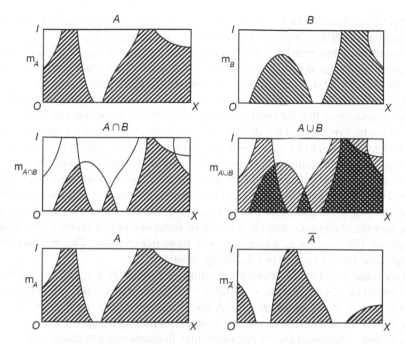

Fig. 19.1 Extended Venn diagrams for two fuzzy sets, A and B, with membership function m(x) and the application of set-theoretic operators union, intersection and complement

Fig. 19.2 Fuzzy cube with the point representing the joint values for three objects, depicted in two dimensions

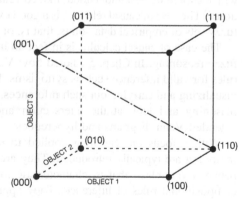

of membership values for n objects (Fig. 19.2). The corners represent the classical sets with membership values 1 or 0.

The diagrams attached to the symbolic formalism have some of the cognitive uses of diagrams I have listed in Chap. 5. They contribute an intuitive grasp of symbolic relations that helps relate them visually to more familiar ones in classical set theory. They also allow for basic level of computation and inference, at least to some approximation. This is a consequence of the fact that the mathematical structure of fuzzy set theory and its application derive from the conceptual and empirical applications of basic set-theoretic concepts and relations, especially the set-theoretic operators whose application are represented in Fig. 5.2.

Determining membership functions plays a key part in the application of models for processing data as a regulated activity. The constraints involve desired targets and the connecting rules. A variety of computational algorithms do this, for instance neural networks.[4] Configurations of input-output units with weights associated with each connection so that the configuration yields a final output that then is compared with a desired output and the difference used through propagation rules to update the weights in order to let subsequent iterations successively approximate the target. Almost like engineering blueprints, here diagrams typically guide the design, exploration, implementation and communication of this kind of learning algorithm (Fig. 19.3). They become a visual environment and a tool for performing different kinds of tasks. Diagrams, in that sense, may resemble maps. In this case, they represent the configuration of the network of input and output layers with symbolic labels for different roles, parameters and numerical values. The representing is inseparable from a context of interacting practices.

One basic set-theoretic relation is inclusion, defined in terms of the arithmetic ordering of membership degrees: subset A is included in subset B if and only if for all members y of B, all members x of A have degrees $m(x) \leq m(y)$. Now, because inclusion is the relation out of which a number of criteria of causality are defined, the distribution of empirical data in Venn and cube diagrams suggests causal relations and enables causal predictions. At least it does so qualitatively and approximately. Another limitation derives from the limited conceptual adequacy of the causal criteria, which should be best understood, like correlations, as empirical signs of causal relations.[5] The case of causal relations is a good example of the role of diagrams depicting fuzzy sets of empirical data points, that is, of thinking with visualizations of data.

The case of causal calculus is in fact an instance of type (2), the visualization of fuzzy reasoning. In Chap. 5 I noted how Venn diagrams enable the application of rules for valid inference such as syllogisms. In fuzzy set theory, to use diagrams for visualizing and carrying out such inferences, we must follow the example of causal reasoning and look at the intersection and subsethood relations visualized by extended Venn diagrams and hypercubes.

With a focus on reasoning applied to empirical facts, more general rules of reasoning are typically introduced. They are generalizations of functional relations between variables—that is, characteristic functions between classical sets—called compositional rules of inference. They approximate, or generalize, classical rules such as *modus tollens*, *modus ponens* and the *hypothetical syllogism*. As in the case of causal reasoning, the underlying assumption is that connected sets model logical implications extensionally—materially—so that the classical characteristic function $y = f(x)$ represents the implication, if x, then y, or if x = a, then y = b. The compositional rules map fuzzy sets and their quantitative membership functions with

---

[4]See Kosko [9], Bishop [10].

[5]I have criticized the conceptual adequacy and application of several causal criteria in Cat [11].

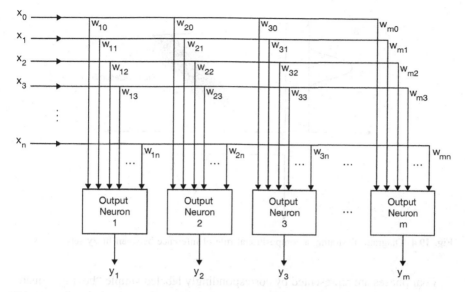

Fig. 19.3 Diagram illustrating an example of a neural network

the help of set-theoretic operators—typically combinations of maximum and minimum operations—that model different implication relations.

   This strategy for adapting logical reasoning to fuzzy set theory borrows the Boolean approach to logical connectives: it includes the familiar rules for transforming conditionals and rules of inference into connectives such as conjunction and disjunction that then are interpreted in terms of set-theoretic relations such as union and intersection and assigned quantitative measures. The result interprets multivalued logics as a model of approximate reasoning and of the formal and material technology of fuzzy control.

   Like classical functions, the new inferences can be visualized by means of diagrams that replace the line, where each point coordinates two values, with a broader two-dimensional region (Fig. 19.4).[6]

   Another example of visualization of fuzzy concepts places it within a diagram representing the architecture of a hybrid rough-fuzzy scheme of image analysis. Like the neural network diagram, the blueprint for this computational architecture works as a dynamical map of a cognitive process. It consists in a series of boxes representing—also instructing—practices applying different formal techniques in a sequence represented by connecting arrows. The scheme's input is a vague image and its output is a measure of categorization accuracy.[7]

---

[6]From the graphic standpoint, design constraints such as the medium of display and cognitive constraints require that the one-dimensional curve is technically a thin area. Geometry is one thing, graphic depiction another; they constitute different kinds of constrained practices applied in different kinds of contexts.

[7]Hassanien et al. [20].

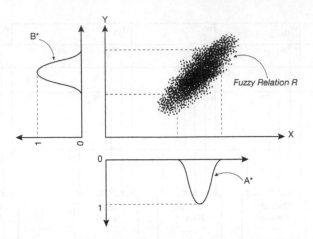

Fig. 19.4 Diagram illustrating a compositional rule of inference between fuzzy sets

Four phases are represented by correspondingly labeled simple "boxing" squares —pre-processing, clustering and feature extraction, rough sets and classifier. Of the four, fuzzy techniques are associated with the two steps of pre-processing, represented by two sub-squares included as in a Venn diagram: intensity adjustment— using a fuzzy histogram hyperbolization algorithm—and enhanced images—using a modified fuzzy c-mean clustering algorithm. The labels for the fuzzy algorithms appear in side "boxes."

19.2 Visual Thinking from Perception to Pictures

The last example of visualization may be categorized also as an instance of type (4). It represents fuzzy models for representing and treating fuzzy pictures insofar as the pictures are fuzzy at the intrinsic level—categorization within a perceptual classificatory scheme—and the extrinsic level—interpretation within a more abstract classificatory scheme for the purpose at hand, e.g., facial recognition, botanical classification, meteorological prediction or medical diagnosis.

I take categorization to be a degree-zero of visual thinking. This places perception in a relation of interdependence with thinking; as I have discussed, representation and reasoning support each other, representation may be extended to include perception and reasoning may be extended to a more inclusive set of activities and processes vaguely called thinking. Other activities include reasoning, computation, prediction, conjecturing, credentialing (with signs and evidential support), learning and problem-solving. Through them we extend the credibility and amount of information. So, how do pictures and their fuzziness enter these activities? How can fuzzy set theory accommodate fuzzy visual thinking?

As I have mentioned, perception has been modeled in terms of thinking, e.g., computational models in the tradition of Marr's work. Accordingly, much work on fuzzy perception is focused on modeling pattern recognition, especially with neural networks and similarity operators.[8] Tracking mental images, tracking perception and tracking pictures may be phenomenologically and cognitively equivalent tasks that are part of visual thinking. The fuzzy case is no different. And in this sense types (3) and (5) may be related. But this is not what I mean by visual thinking of type (4) with fuzzy pictures whose perception isn't itself fuzzy; unless we understand fuzzy recognition as thinking concerned with the perception of a pattern in a picture and we assume that seeing a fuzzy pattern and seeing a sharp pattern fuzzily are more than phenomenologically indistinguishable.

The specific type of case in type (4) I want to emphasize is the kind of visual thinking that depends on visual interaction with the environment in which fuzzy representations occur and may be subject to fuzzy set modeling. This eliminates the cases of exclusive focus on internal images in conditions of "introspection."

When thinking extends beyond categorization into the kinds of activities mentioned above in visual interaction with the environment, it applies to mental images and pictures (sources of mental images) differently, that is, as separate but related types of representations that involve different but related kinds of practices and constraints, especially if we include the domain of scientific practices.

In earlier chapters I have examined the special part of the perceptual environment that pictures are. I have presented evidence from perception-based models of pictures; they argue for the role of opaque contexts in which we produce, interpret and use pictures. The perceptual core of the accounts emphasizes the special relation between seeing and a material medium.

Unlike the case of images as "representations" in perception, in pictures perception is characterized by a dual experience of seeing the medium and seeing in the medium, seeing a mark and seeing in the mark. Seeing-in can be interpreted in terms of Gestalt-like seeing-as, perceptual categorization that identifies intrinsic content and links it with extrinsic content—we see a mark of paint on a canvas, for instance, as a human face or a symbol. But such cognitive activities are embedded in the application of systems of conventions, design styles, display media, etc. It is not just the symbolic structure of diagrams, with their symbolic conventions and rules; but even the perceptual realism that reproduces the subjective, perspectival standard of accuracy requires skill and learning.

Perceptual realism reproduces some of the effects that we associate with environmental cues such as perspective, even blur that triggers projections of categories. But this means that by continuity with perception, it is constrained by some of the same effects while it exploits them. Diagrams themselves, like any picture, cannot escape the scope of perception. The contrast between sharp graphic display of geometric features and rough and blurry photographic realism is relative. Visual sharpness and fuzziness are relative. In a given context, our cognitive conditions

---

[8]For an introduction, see Klir and Yuan [7, Chap. 13].

and the material properties of the medium yield a standard of sharpness against which we draw contrasts. In fact, this is the sort of contrast that allows us to distinguish between eyesight unreliability and a fuzzy picture surrounded by sharper features and vice versa (Figs. 7.1 and 7.3).

19.3   Reasoning with Visual Analogies and Pictorial Evidence

The use of fuzzy pictures enters visual reasoning in different roles such as providing visual evidence and reasoning by visual analogy. Unlike mental images from individual perception, the material objectivity of pictures allows for an empirical sort of intersubjectivity; this in turn supports the empirical application to the content of the picture of methodologically objective reasoning procedures.

Analogy plays a role in the application of fuzzy, rough and near sets. They all include similarity criteria that require co-categorization, that is, sharing or being indistinguishable in regards some feature. Note that the emphasis on the relation of indistinguishability brings out the cognitive focus of the formal treatment; it models a contextual cognitive state.

Specific similarity measures (and associated operators) quantify degrees of similarity of the perceptual objects relevant to a prototype—by deformation—or metric of relative similarity in the context of a cluster of objects or a cluster of properties. We can compare two objects by comparing not just their degree of membership in relation to a feature or categorization, but in relation to a set of features. Like inclusion, similarity measures are defined in terms of basic set-theoretic relations union, intersection and equality. In all cases they are measures of co-categorization.[9] As a relation, similarity is reflexive and symmetric but generally not transitive. Fuzzy measures can qualify the three properties. The relation and its measure are as contextual as in the determination of the single case of degree of membership (and classification more generally). As perceptual objects, the approach applies to pictures within the same range of possible categorizations.

Now, the role of similarity here is not one contribution to the picture's role as a representation. Analogy is not analogical reasoning. Reasoning by analogy is a form of making inductive inferences from a set of features to a larger set. In doing so, the similarity relation acts as a vehicle for the communication of intrinsic or extrinsic content; and for this the relation must be partly transitive, a feature of fuzzy measures.

What makes it an inference? It is the rule that applies the conditional that helps track the recognition of a shared feature and its association with another. The linguistic rule is assigned a quantified set-theoretic representation in terms of basic relations such as the compositional rule of inference, above visualized in

---

[9]Dubois and Prade [12, Chap. 3].

Fig. 19.2. If the quantitative rule of inference is satisfied by a conjunction of propositions from one set and a proposition from another, we have a consequence relation between the first—the set of premises—and the second—the conclusion, or consequence. If the relation is reversed, the second constitutes a hypothesis. If they do not satisfy the relation, the second is a speculative conjecture.

But in the case of analogy, the best we can do is to introduce a conditional based on the additional application of a quantitative constraint on the inference, that a degree of similarity or indistinguishability meets a certain critical minimum. Then, with the aid of the rule, we can derive measures the uncertainty from antecedent measures of similarity. Then we can perform some analogical "reasoning" by grafting the analogy onto any of the three relations of approximate reasoning, above. If the conjunction of premises is analogous enough to the conjunction in the first set, we can use the inference rule to generate additional consequences, hypotheses and speculative conjectures.

In the strongest case, any original uncertainty in the premises or the conclusion is compounded by the degree of similarity. In general, then, analogical "inference" plays a heuristic role, as a source of hypothetical and speculative conjectures for further testing. It is a thinking procedure rather than a type of strict reasoning. In the visual case, we do not connect propositions, but compare categorizations—between sets of features—so that next we associate with one the set of features of one picture a property from the analogous set as the cognitive outcome—by way of approximate conclusion.

For instance, the similarity between a diagram representing the pattern of evolution of electric activity in a brain or a heart and another may suggest that also the first pattern, like the second, indicates a certain physical condition and helps expect a certain evolution, by way of prediction. One property is the basis for inferring another by association and extending our information about the system in question. By itself the argument is based on a set of representations and joint categorizations linked to specific contexts. Fuzziness enters the picture with categorization, either at the intrinsic level or the extrinsic, or both. Those are the empirical "premises" that the approximate set-theoretic standard of consequence connects.

We can identify another mode of thinking by reversing the order of the categorizations. Give a hypothesis, we can provide evidence in support of its credibility by pointing to an analogy by way of a further consequence. We believe more strongly that something is an organism of a particular kind and it will act in a particular way as a result because it resembles something else of that kind, with a shared feature set. But this looks like the basic case of categorization and classification by degree of membership, or at least it includes it; representation often relies on reasoning...by analogy. Typically, then, this sort of inference will increase the plausibility measure of the conjecture with more than one analogy, as in argument from robustness.

The methodological significance of the contrast with the heuristic function depends on whether and how we draw the logical asymmetry between evidence and symptoms—signs or indices. A sign or symptom of a disease might be an index caused by the underlying condition or at least be empirically correlated with it; it is subsequent test results that play the role of evidence in favor of the diagnostic conjecture. We can conclude that it's not just the epistemic and practical matter of finding objective formal criteria—quantitative or logical—for changing our beliefs, the purpose of evidence theory; it is also a matter of the methodological ordering of cognitive activities.

This points to a more general issue about visual reasoning, which is the practice of arguing with visual evidence, not proof, not a formal argument based on quantitative information and computations. In the pictorial context, the set-theoretic relations that formalize analogy and inference apply to objects of perception, not linguistic propositions. Their spatial, intrinsic properties are maximally isomorphic to the properties of the external target system, similarly dense, and caused by them. This fact raises two familiar and connected problems about the relation between pictorial depiction and linguistic description: what sort of relation between descriptions and depictions transcribes the visual representation into a proposition standing for a pictorial belief that sustains inferences? What sort of relation between a picture and a hypothesis is one of verification, evidence or support?

I have mentioned the transcription strategy. If the picture is a representation of numerical data, the transcription will involve a substitution the original data for its graphic display. If the original data isn't independently recoverable, we will face the uncertainty-filled task of identifying the quantitative information from the geometric visual one, without a systematic mapping. I have also noted that useful pictures, especially in the sciences, but not exclusively, are hybrid pictures, analogical and symbolic, dense and discrete, indexical and conventional. The pictorial, spatial properties help identity more abstract, extrinsic relations between the symbolic data.[10]

An alternative consists in mapping fuzzy-looking displays of data onto fuzzy diagrams, so that the fuzzy-looking visualization can be either transcribed into quantitative fuzzy relations or used for visual computations, "inferences" and other cognitive tasks. But, without membership degrees, in both cases we have at best a qualitative mapping onto, for instance, fuzzy Venn diagrams or hypercubes. Any rules for graphic transformation will depend on the way the graphic display visualizes categorization, for instance, pie charts, graphs and Venn diagrams.[11]

Visual data is not reducible to data visualizations. In the case of analogical representations such as photographs, the transcription challenge is compounded by the I determinacy of the background IC–EC link, the sort of rule that is often juxtaposed in linguistic expression, mediated by symbolic code superimposed on the picture. Schematic diagrams and maps present the same difficulty. On the linguistic standard, the reasoning will involve only the linguistic descriptions of the visual data. And their objectivity will require consensus about their adequacy. This strategy will preserve the notion of degree of truth in the use of fuzzy predicates at

---

[10]Perini makes this point in Perini [3] about tables.

[11]One kind of mapping from fuzzy set diagrams to classic Venn diagrams relies on precisification techniques such as so-called alpha cuts.

the core of set-theoretic treatments, although we have no guarantee that any accepted linguistic and pictorial contents will match.

Still, the adequacy of description depends on the interaction between description and depiction: the pictorial content, at the intrinsic level, rests on causal relation with the spatial structure that it represents; the extrinsic identification and selection of relevant features is determined by the description and the background knowledge that supports it.[12] Unlike diagrams, realistic pictures carry evidence because of the causal, not logical, relation to their content. The epistemic value of the partial truth of propositions depends on representation of the fuzzy pictures, their form and causal indexicality.

On my approach, the hybridity and the compatibility with the set-theoretic formalism resides in the basic role of categorization. With our without transcription, it is the association between membership degrees of the different shared categorizations—intrinsic, extrinsic, external—that enables the application of rules of inference, even without truth-valued propositions. In fact, the only truth assumed by the application of fuzzy set theory does not require any strict semantic or ontic models of facts about "real" properties. Whatever might trigger and implement recognition, degrees of categorization may be interpreted as degrees of property instantiation, but it is only an interpretive choice.

> As set theory models cognitive practice, categorization provides both the conceptual and causal link to the empirical world of objects we classifies and the context of practices within which we do so. The interaction is more encompassing and complex than the indexical link. It involves recourse to the exercise of perception and memory, and reference to prototypical individuals among others. Since the relevance and evidential role of experience is not merely causal but informed by a context of shared goals, skills and standards, the categorization connection plays a more limited but also a more reliable and persuasive evidential role with its contribution to argument, formulated linguistically or symbolically and at least by a given set-theoretic standard of inference.

19.4   Problem-Solving in and with Pictures

If we understand categorization as in the context of image processing, as an activity of extraction of uncertain features and information, we can easily accommodate the role of images as sources of information for computation, inference and, more generally, problem-solving. Problem-solving is probably the most generic form of visual thinking, and also a particularly valued test in support of the cognitive role of mental imagery[13]: the processing of visual information in the performance of a task relative to a set goal and a set of constraints that are both limiting and enabling. For

---

[12]In Perini [3] Perini defends the hybrid interaction for images in general.
[13]Kosslyn [19].

the purpose of this discussion, we can distinguish two main kinds of problems defined and solved in visual, spatial terms:

- *Visual problems*: problems defined in visual terms with tasks and goals to be engaged in relation to a visual environment, especially a picture.
- *Visual problem-equivalents*: problems in which the visual features are singled out as visual illustrations of symbolic terms in which the problem is originally defined—visual equivalents must include visual expressions of the relevant form of the solution, the task, etc. and their components.

Visual equivalents, a subgroup of which will be analogs, are the most common kind. They provide an effective and convenient tool for solving problems within an accepted margin of approximation. Examples are mathematical computations formulated in arithmetic terms performed in geometrical terms, medical uses of X-ray images to plan the performance of a surgical procedure, and the many uses of maps to help with practical tasks requiring information about distance and travel routes.

The relevant pictures will be instrumental in enabling thinking by enabling the representation of problems and their solutions. To solve the problem is to be able to identify the representation of the terms of a problem and use them to identify a representation of a solution. However, the use of maps and photographs as useful representations is not a matter of simple analogy, at least for two reasons. First, the units of equivalence, in particular the units of analogical representation, are not uniquely determined by the units into which we can analyze the terms of problem and its possible solutions. Ideally, the picture must represent all possible solutions. What counts as a representation of a solution? If we identify each solution with a series of tasks, the picture must represent the necessary spatial features. But the design constraints on the picture might prevent the partial—insufficient—or complete—insufficient—representation of all solutions (by some standard of completeness or sufficiency of analogy). An X-ray, a photograph or a map might not be functionally complete; hence the coordination with verbal reports might be supplemented with physical gestures. Similarly, there is no guarantee that every possible part of the picture will correspond to—represent—to a spatial part of the unit task. As a case of representation, the mapping violates the Sober–Fodor principle of compositionality. Second, since the process of recognition that we represent in terms of categorization, especially for the purpose of set-theoretic treatment, involves a number of interpretive skills, standards, background assumptions, conventions, etc. The representation of each categorization might be mediated by a diagrammatic element such as an arrow that specifies key dynamical features of the tasks involved.

The sources of fuzziness in all these kinds of images for the purposes at hand are multiple. To represent them and implement them in technological contexts, the application of set theory begins with fuzzy categorization. But I have stressed that fuzziness might concern the extrinsic as well as the intrinsic content, so that given a certain standard of perceptual description, nothing in the picture looks fuzzy. This might concern the depiction of terms of the problem as well as terms of possible solutions. With both kinds of contents we can model, design and implement representational and problem-solving tasks.

What is significant in formal reasoning models is the role of rule-based algorithms. If-then rules are introduced for the sake of the extraction of categorized knowledge from a data set, in the case of images, perceptual data or information. In general, then, this knowledge and the basis for the different forms of visual thinking take the form of classification: 'knowledge is primarily defined by the ability of the system to classify data or objects'[14] and 'inference (reasoning) is understood as the final unique assignment of an object under consideration to a specified class.'[15] Reasoning in this sense provides the links between observation, comparison and recognition or classification.[16] The key step is the generation of an inference rule (if-then conditions) that assigns a class (so-called system learning). Typical inputs (antecedents) are either weighted vectors of feature values (with associated fuzzy membership grades) or fuzzy clustering with standards. We can connect intrinsic and extrinsic contents in different ways considering differences in their respective degrees of fuzziness. Still, the successful design and application of the rules is part of a complex modeling activity representing similarly complex and contextual cognitive activities.

19.5 Final Remarks

Again, when we focus on pictures we encounter a familiar foundational challenge. Representation is represented logically in terms of the analysis of pictorial content into units of categorization. But can reasoning with fuzzy pictorial representations be equally performed and represented? Representation-based reasoning and thinking in general are a calculus of elementary or composed categorizations. The only semantic relation is pictorial accuracy and the only empirical truth is truth of categorization. If pictorial accuracy or fit cannot be identified with linguistic truth, fuzzy pictorial inference must be represented within a calculus with inference rules without (partial) truth conditions, only fuzzy representation—that is, categorization—conditions.

Are there linguistic alternatives that preserve the semantic foundation of the theory? If we wish to avoid the radical foundational alternative I have suggested, two strategies are these: (1) direct pictorial calculus based on what I have called PR depiction relations as instances of categorization or set-theoretic relations between them in intrinsic and extrinsic contents; (2) indirect linguistic transcription based on transcription rules or constraints that yield forms of predication supporting truth valued propositions, including representation reports. I have also discussed the complexity of hybrid interactions. All these strategies are based on categorization or

---

[14]Mushrif and Ray [13, 10.2].

[15]Tarnawski et al. [14, 6.2].

[16]Peters and Pal [15].

instantiation rather than truth. And the truth content they support might not exhausted the pictorial content. Nevertheless, both might involve a sufficiently adequate notion of accuracy or fit to make apply fuzzy set theory and make sense of it too.

Cases of composition are forms of content development and include the two scenarios of extrinsic depiction relations presented above between the image and its object, the object of which it is an image, as described in the extrinsic content. As a matter of degree, the fuzzy image in a picture is, say, (partly) face-like by virtue of a coordinating IC–EC rule, of (partly) being of a particular shape, involving a general distribution of specific shapes, color, etc. IC–EC rules, whether empirical hypotheses, theoretical hypotheses or conventions, play the role of inference or computational rules.

Now, the image represents a face if the categorizations instantiated by the image as sufficiently (partly) shared by a particular face or a class of face-like images (the second scenario is relevant especially to the case of abstract or fictional entities such as a type or a character). Then we can add the auxiliary rule that identifies a feature in the image as an image of something with another feature of the object in question, extending the extrinsic content, for example, a distribution of color spots on the otherwise colored image of the face, or a brain scan, may represent in turn a particular type of structure, entity, property or condition of the face or brain thus represented such as a skin infection, a kind of brain activity, or a tumor.

Neither case of depiction by the picture, or by a class of similar pictures, is a case of a semantic relation of (partial) truth, (partial) truth value or degree of truth. When we say that the brain picture vaguely represents the brain tumor if a particular brain instantiates or can be categorized similarly enough (besides a indexical causal relation, etc.), nor is the relation of representation between two cases of instantiation or categorization a relation of (partial) truth—between two relations of (partial) truth, (partial) truth values or degrees of truth.

The rules of inference that are interpreted by particular hypothetical, empirical or conventional IC–EC relations cannot be defined in terms of degrees of truth. But they can be given a formal set-theoretic structure relating membership degrees by analogy with the many criteria for the conditional operation alone familiar in fuzzy logic to calculate or test degrees of membership.[17] The degrees may differ between the different objective categorization or instantiation events involved (the picture, the brain, etc.), not only because of the failure of PIC–PEC rule, also because of the possibility that only the intrinsic content or the extrinsic content may be fuzzy. The rules of inference just do not provide a truth calculus. Operations cannot be defined on strict truth tables. Instead, we can aim at a set of conditions in terms of a weaker form of correspondence or relation satisfying the semantic function of truth for pictures.

Nevertheless, we can recover the symbolic version of fuzzy reasoning in a derivative way. The different categorizations can be modeled in terms of degrees of truth by superimposing a linguistic structure, namely, by attaching a fuzzy predicative proposition working as a representation report. The representation report, in

---

[17]Dubois and Prade [12], Novak et al. [16], Klir and Yuan [7], Lawry [17].

turn, may itself be decomposed into—or derived from—an elementary fuzzy categorization report on each object, the picture and its target. The case of fuzzy diagrams, already a hybrid of pictorial and symbolic, analogical and digital, will be closer to the symbolic fuzzy calculus. Still, the challenge consists in identifying and modeling a variety of reasoning activities involving fuzzy diagrams (identifying realistic conditions where diagrams may present a fuzzy IC or a fuzzy EC) and fuzzy pictures.

I have pointed to the relation between perception of pictures and perception trying to locate the role of pictures among the role of images when thinking about representation in depiction and in perception. The emphasis on representation may be helpful to make sense of certain practices and to carry them out; but it may also be misleading, making it the fundamental role of categorization in its context. Visual thinking is cognitively similar to a visual process; as such, it is similar to different forms of empirical inquiry in the perceptual environment in which we are embedded and act. In that environment pictures enter in different ways, in the role of perceptual sources of information that we associate with the practices of representation, for instance, visual representations of quantitative data. Examples are modes of pattern recognition in data plots with explanatory or predictive value or with evidentiary value as a matter of qualitative fit. The context of categorization is more complex.

The equivalence can be strengthened by the assumption behind the analytical set-theoretic approach to images and image processing, that an image is a set of visual data points. Reasoning with images as sets of visual data points will suggest new heuristics for reasoning with data visualizations. To the extend than diagrams, for instance, provide efficient tools for processing information, in lieu of "equivalent" symbolic statements,[18] data images may be taken to provide an efficient too for processing data. Indeed, fuzzy categorization provides an extension of empirical data sets by virtue of its extension of the allowed and meaningful degree of membership.[19] Images, as diagrams, provide mediating models of the data as well as tools for processing them. In conclusion, fuzzy visual reasoning may prove of relevance to the valuable extension of big data sets and their use.

Focusing on pictorial representation in terms of fuzzy categorization allows us to accommodate a more general notion of vague thinking: instead of fuzzy reasoning, fuzzy cognition with pictorial representations. Now a picture's content could work as the input in different kinds of rules for carrying out instructions serving different but related means and purposes: inference as a truth-value-based rule, inference without truth, computation without inference, problem solving without inference or computation, processing and decision-making.

Does fuzzy set theory require truth? From the semantic point of view, to associate the precision of a predicate to its place in a (completely) true proposition is an interpretation of the empirical application of the formalism to model natural

---

[18]Larkin and Simon [1].
[19]Ragin [18], Cat [11].

linguistic assertion. As a matter of formal virtue, the consistency of the formalism is independent of the application of the semantic criterion. As a matter of empirical application, the success of the formalism is independent of the application of the semantic criterion to any assertion; empirical claims are models or hypotheses accepted on methodological grounds. Truth is less relevant to scientific practice than acceptance on grounds for revisability and reliability. From a logical point of view, the validity of fuzzy or approximate rules of reasoning has two sources: a formal constraint that recovers the form of classical logical inferences and the empirical reliability of premises and conclusions as categorization tasks. The rules are, as in so-called quantum logic, empirical hypotheses. Truth in the strict linguistic sense discussed in Part 1 is an additional semantic constraint with limited application to pictorial accuracy. Articulating the form of the rules that serve each task is both an empirical and a contextual matter of formal and practical control.

Finally, let me recall the radical alternative I have merely suggested, that consists in denying the assumption held all along, that the semantic foundation of fuzzy set theory is a strictly linguistic truth. We may weaken this standard, for instance, epistemically, in the sensible scientific spirit of reasonable acceptance. If there is such a thing as a scientific standard of truth, it might be close to this; the rest is a metaphysical hypothesis. Instead, we may reject it altogether. Then we can assume vagueness and inference without truth. The question of validity of rules becomes a matter of objectively reliable categorization.

References

1. Larkin, J. H., & Simon, H. A. (1987). Why a diagram is (sometimes) worth ten thousand words. *Cognitive Science, 11*, 65–99.
2. Perini, L. (2005). The truth in pictures. *Philosophy of Science, 72*, 262–285.
3. Perini, L. (2005). Visual representations and confirmation. *Philosophy of Science, 72*, 913–926.
4. Shin, S.-J. (2012). The forgotten individual: diagrammatic reasoning in mathematics. *Synthese, 186*, 149–168.
5. Goodwin, W. (2009). Visual representation in science. *Philosophy of Science, 76*, 372–390.
6. Blackwell, A. F. (2001). Introduction. Thinking with diagrams. *Artificial Intelligence Review, 15*, 1–3.
7. Klir, G. J., & Yuan, B. (1995). *Fuzzy sets and fuzzy logic. Theory and Applications.* Upper Saddle River, NJ: Prentice Hall.
8. Zadeh, L. A. (1965). Fuzzy sets. *Information and Control, 201*, 240–256.
9. Kosko, B. (1992). *Neural networks and fuzzy systems: A dynamical systems approach to machine learning.* Englewood Cliffs, NJ: Prentice Hall.
10. Bishop, C. M. (2006). *Pattern recognition and machine learning.* New York: Springer.
11. Cat, J. (2006). On fuzzy empiricism and fuzzy-set models of causality: What is all the fuzz about? *Philosophy of Science, 73*(1), 26–41.
12. Dubois, D., & Prade, H. (1980). *Fuzzy sets and systems. Theory and applications.* New York: Academic Press.
13. Mushrif, M. M., & Ray, A. K. (2010). Image segmentation: A rough-set theoretic approach. In Pal and Peters (pp. 10.1–1.15).

14. Tarnawski, W., et al. (2010). Applications of fuzzy rule-based systems in medical image understanding. In S. K. Pal & J. F. Peters (pp. 6.1–6.31).
15. Peters, J. F., & Pal, S. K. (2010). Cantor, fuzzy, near, and rough sets in image analysis. In J. F. Pal & S. K. Peters (Eds.), *Rough Fuzzy Image Analysis. Foundations and Methodologies* (pp. 1–15). Boca Raton, FL: CRC Press.
16. Novak et al. (1992). *Fuzzy logic and decision-making*. Dordrecht: Reidel.
17. Lawry, J. (2006). *Modelling and reasoning with vague concepts*. New York: Springer.
18. Ragin, Ch. (2000). *Fuzzy-set social science*. Chicago: University of Chicago Press.
19. Kosslyn, S. M. (1994). *Image and brain*. Cambridge, MA: MIT Press.
20. Hassanien, A. E., Al-Qaheri, H., & Abraham, A. (2010). Rough hybrid scheme: An application to breast cancer imaging. In S.K. Pal & J.F. Peters (Eds.), *Rough Fuzzy Image Analysis. Foundations and Methodologies* (pp. 1–5). Boca Raton, FL: CRC Press.

Chapter 20
Pictorial Approximation: Pictorial Accuracy, Vagueness and Fuzziness

In this chapter I relate the ideas of fuzziness and accuracy of pictorial depiction to the complexity of approximation. Attention to approximation contributes to my focus on the application of fuzzy set theory as a contextual practice, bringing out some of its aspects as an expression of the practice of approximating. Fuzziness in pictures, then, adds another dimension of approximation; and vice versa, approximation adds another dimension to the analysis of fuzziness. In particular, it contributes a helpful way to understand the contrast between inaccuracy and imprecision.

Representations often get their value from their approximate nature.[1] Approximation suggests inaccuracy, incompleteness, difference, deviation and distance. But approximation also suggests positive cognitive value, in much the same way, I have noted, visual fuzziness does. It is inaccuracy that gives a basic sense of content.[2] In addition, the notion of approximation articulates a difference between inaccuracy and falsehood that rejects strict error or falsehood and the binary dichotomy that enables it as the sole alternative to a single form of success.

For all the value put on ideals of rigor and accuracy, the real practice of science and engineering is effectively well adjusted to idealization, simplification and compromise as valuable features of techniques and results, both in practical and theoretical terms.[3] As a result it also provides a measure of sufficiency or compromise, and a target goal that reveals its role as a condition of objectivity and rationality when assessing representational progress.[4] Whether attainable or not, a sense of approximation might serve a dual function, as a limit that provides a sense of direction and destination and as a reference point that provides a sense of

---

[1]To repeat the qualification, I have already insisted on the plurality of uses and values of images beyond representation and its own uses.
[2]Siegel [1].
[3]Feyerabend [2], Kline [3].
[4]Bachelard [4].

© Springer International Publishing AG 2017
J. Cat, *Fuzzy Pictures as Philosophical Problem and Scientific Practice*,
Studies in Fuzziness and Soft Computing 348,
DOI 10.1007/978-3-319-47190-7_20

distance. In this way it enables the assessment of representations, albeit contextually, with the desirable result of making possible judgments and decisions.

From that standpoint, nothing prevents applying the notion approximation to pictures as pictorial representations, even if representation isn't reduced to linguistic truth. With the particular focus I have selected, approximate representation is a matter of categorization in relation to each of the two main kinds of pictorial contents, intrinsic and extrinsic. Now, the question is whether vague pictures too can be considered approximate. My answer is that pictures not only can represent approximately; to repeat, fuzziness in pictures adds another dimension of approximation; and vice versa, approximation adds another dimension to the analysis of fuzziness.

The practice of approximation may be characterized in terms of the following five related features: *structure, metric, target, standard* and *contexts of significance*.[5]

(a)   *Structure*

From the point of view of mathematical formalism, the bare-bone interpretation of approximation is purely formal. In structural terms, it is a relation internal to the symbolic structures that characterize the approximating concept, and this relation is determined by rules and standards of the particular formalism. On this basis one can then introduce approximative forms of reasoning and computation.

(b)   *Metric: formal and empirical*

The metric dimension of approximation is the structural feature that enables a particular determination of proximity on a distance scale. Necessary conditions include a *partial ordering* provided by the structure, *additivity*, *transitivity* and the specific structure of a *measure* function will contribute towards a metric structure with a measure of distance. In nesting structures or recursive functions, the ordering gets a cardinal measure in terms of the number of iterations or level of nesting. The property has different expressions in the methodology of empirical sciences, including in more specific analysis of theories of empirical measurement.[6]

(c)   *Target: formal and empirical*

Attention to proximity, or closeness, provides a yet more specific consideration of generic distance. Proximity is not distance. As a representation of distance, proximity is a *symmetric* relation involving each term in the series. But the formulation and the interpretation of approximate claims typically assume an added asymmetry linked to the notion of deviation. It is related to at least two connected constraints, namely, restricted *range* and *centeredness*. On this basis approximation acquires an asymmetric connotation characterized by practical dimensions of dynamics and value.

The implied restriction on the relevant range fixes the bounds of proximity but it is typically left unspecified. The assumption is that, while the domain of proximity may be in principle indefinitely large, in practice we find a smaller interval of values

---

[5]I introduce this account in Cat [5].

[6]An encyclopedia of formal conditions for measurement is Krantz et al. [6], vol. 1.

or a subset of terms—e.g., in an ordered series. This is an expression of a choice informed by a scientific interest within the practical context of engaging the formalism—that is, exploring it, developing it and applying it.

The series is presented relative to at least one fixed reference; typically the approximation interval will involve two terms, even if one is infinity. Indeed, all partial approximations, embedded in a metric ordering or structure, share an effective reference point, which may be a value of a variable, a form of a function, etc. This reference is what I call the *target* of approximation; it constitutes the standard of exactness and introduces an *asymmetry* in the relation of approximation. The role of this reference in definition or application is to impose a preferred order and, relative to it, determines the intended measure of degree in a comparative form. For instance, some relevant value is approximately 0, or it tends to 0, or a linear term is first in a polynomial series expansion of a function at a point approximating to that order an exact solution, etc.

When a set of values or functions share a target with the same measure of proximity to it and standing in a symmetrical, inverse relation we might call the target the *center* of that set. The target plays the role of center of the approximation range.

In the context of an empirical application, it is sometimes a symmetric range around a designated center that is of relevance. In the case of measurement, we find the assumption that margins of measurement error, or degrees of accuracy, form equivalence classes relative to shared empirical target values.

(d) Standard: formal and empirical

Approximation relations receive different use, value and meaning through their relation to different external standards and associated contexts of significance, that is, in relation to varying kinds of situations and perspectives.

A standard, a baseline or threshold, is typically related to a context of significance as its condition of application of (significant) aims and constraints. The relevant perspective will incorporate additional standards or norms that provide the normative source of significance.

(d) Contexts of significance

What are examples of the different contexts of significance in our understanding and use of approximations? Each kind illustrates a perspective, situation and associated values and purposes: categorization, generalization, reduction, scope of validity, precision, accuracy of prediction, evidence, degree of objectivity of properties of a system or of cognitive states, explanation, degree of progress, etc. Each perspective includes corresponding kinds of regulative aims and standards, or conventions, and with these it formulates and engages different sorts of projects and problems.

The specifics of each context sometimes differ by disciplines, sometimes by empirical situations or interests (in some cases connectedly, in others separately). The role of standards is to connect with the context of significance and help apply its terms, norms and goals. It answers questions about why the target reference matters, how far from it makes a significant difference, and what difference it makes. Examples are critical levels as safety standards in connection with setting a level of danger or risk, e.g., levels of radiation, temperature, toxicity, white cells count, public debt or violence.

As a scientific practice, approximation has a number of more specific contexts of significance. Among possible such contexts I distinguish the following[7]:

1. Formal conceptual
2. Formal computational
3. Applied conceptual
4. Experimental
5. Evidentiary
6. Theoretical, interpretive and explanatory
7. Pragmatic
8. Disciplinary
9. Dynamical and historical (as practice and directional process)

Iterative processes illustrate contexts 1, 2, 3, 4 and 9. Neural network models go through dynamical cycles of revision of weights and outputs until they converge on a stable output value. As a model of a learning process, we hope the output value to approximate as closely as possible to a target value. The history of much design and calibration of measurement instruments and the use of hypotheses to generate data in support of newer and contradictory hypotheses also fits this pattern of evolution. Dynamic approximation is a process and a practice.

From the general standpoint of representation without truth, degrees of pictorial accuracy satisfy the model of approximation as a cardinality measure over sets of properties. The judgment of approximation applies to the different sets of categorizations and associated properties that instantiate them insofar as they correspond to the different kinds of contents I have claimed a picture can bear: *intrinsic*, *extrinsic*, *external* (instantiated) and *incidental* (a set of properties or entities represented according to a given criterion but devoid of access, relevance or acknowledgment according to constraints of the given context. In particular, different criteria for identifying or producing representations such as similarity and indexical causal link will admit of different incidental contents). As a matter of content, considerations of approximation may be said to apply to perceptual representation as well as to externalized depiction.[8]

When representation depends on shared categorization, or, equivalently, on instantiation of the same property, by the picture and by its object (the target system), we typically think of similarity, whether perceived or not.[9] In that case, approximate representation may take the form of the number of such co-exemplified properties, and, therefore, of a degree of similarity (perceived or not).

[7]See Cat [5, 7] for details.
[8]On perception see Siegel [1] and Falguera and Peleteiro [8].
[9]For a defense of the more restrictive criterion, see Hopkins [9].

Naturally, the measure of approximation suggests a measure of degree of completeness or closeness to identity. The difficulty lies in establishing the metric based on the standard of completeness or identity. The sense of approximation derives more from a measurable increase and a limit to infinity than from the indeterminate numerical difference between similarity and identity, completeness and incompleteness, absolute accuracy and inaccuracy. In any given context, we can always apply an effective cognitive standard of identity or completeness in terms of indistinguishability. This is the illusionistic limit that has been sometimes defended as the standard of representation achieved especially in photographic perceptual realism. Short of that standard, the sense of approximate success takes the equally contextual form of sufficient recognition.

As a measure of approximation, a qualitative property count is not sufficient— size, figure contour, color range, brightness, etc. One must add, for instance, considerations of scale as a measure of the degree of resolution or detail; they recall the Sober–Fodor condition on imaging, that parts of the image represent parts of the target system, in our case by categories that their recognized properties exemplify. In this way we may consider a map or a photograph a better visual approximation than another. But the judgment is relative not only to the material constraints on the medium or process of production. It is also relative to the interests and purposes at hand; too much detail or too small a scale might be as irrelevant as too large a scale or too low resolution.

More directly relevant to the case at hand, it is also relative to each individual property or category. Vagueness concerns the individual case. Degree of fuzziness is an approximate measure at the core of vague representation and reasoning. This is compatible with the objective character of vagueness. Since a central theme in this book is the diversity of representations and uses of vagueness or imprecision, I claim that three technical set-theoretic concepts I have presented, fuzziness, roughness and nearness are different types of approximations.

The formal structure of fuzzy set theory around the calculus of degrees of membership fits the standards of approximating practice listed above.[10] The range of membership degree values in the range $[0, 1]$ and the standards instantiating at least full membership, with value 1, in a given context, provide the basis for thinking in terms of approximation. One of the criteria for determining the degree of membership is the degree of similarity to the standard that in a given context stands for degree 1; how similarity is assessed in turn is less rule based, except via clustering methods with their own sources of subjective and pragmatic elements. The similarity approach extends its counterpart for a set of categories and its role in fixing the notion of approximation.

While degrees of membership don't guarantee an approximate semantics of degrees of linguistic truth, they provide a measure of approximate categorization and, derivatively, of properties constructed on its measure. This is the basis for thinking about fuzzy representation as the fuzzy version of the proper function of

---

[10]Cat [5].

pictorial representation as distinct from linguistic truth. On this basis fuzzy depiction qualifies as a form of approximate representation; the conclusion extends to reasoning as either based on or part of the construction of representations.

In the context of empirical application, then, one can make sense of the notion of fuzzy picture (and predicate) as an element of approximate representation and reasoning over en empirical domain. Considered a measure of qualitative accuracy, fuzziness also in pictures shows vagueness to be another dimension of approximation. Like others, also fuzziness has its contextual dimension of significance; I have pointed to the pragmatic elements of application and problem of PIC–PEC coordination. Not all degrees will have the same consequences and fare equally in any given set of constraints and purposes.

I have also mentioned two other formal characterizations of imprecision: roughness and nearness. Recall that according to the rough-set generalization of classical set theory, imprecision is not a matter of boundary vagueness but a matter of indistinguishability between individuals or sets. The focus is on knowledge of the universe of objects through their categorization and classification in terms of a series of features and their values. Each category corresponds to a subset of the universe. In the pictorial domain, the individuals may be visible objects, pixels or data points described in terms of perceptual features. The selected features partition —classify—the universe into basic equivalence classes of indistinguishable objects; then, if a set can be represented as the union of basic sets, it is called exact or definable in basic terms. Notice that exactness is a function of the descriptive imprecision or low power of resolution that classes things together.

Rough sets are the ones that cannot be defined within the inevitably broad brush of the knowledge base; and rather than representing their inexactness in terms of fuzziness of their boundary objects, the theory represents it in terms of a relation of approximation to two exact sets whose objects fit exactly within the building blocks of indiscernible objects. The so-called lower approximation is the union of subsets that are included in the rough set. The upper approximation is the union with a non-empty intersection with the rough set. The boundary region between the two approximations is the representation of the inexactness of the rough set. In the initial formulation of the theory, Pawlak called the boundary region the doubtful region, emphasizing the cognitive aspect of the problem in AI and machine learning.[11] In contemporary hybrid algorithms, the boundary receives the additional description in terms of fuzzy membership values.

Accuracy is measured by the ratio, a, of the size of the lower to the upper approximation extensions, leaving fuzzy membership to model overlapping partitions. Roughness is then defined in terms of inaccuracy (1-a). In line with my discussion above, roughness is an epistemic matter of classification of objects in terms of classifications of information or data, where the mapping onto the range of feature values provides a description. The data may also be the outcome of decisions in the automated application of rules. The rough-set measure of (in)accuracy

---

[11]Pawlak [10].

receives the additional epistemic interpretation as (in)completeness of knowledge of the rough set and, by inclusion, of the universe.

Also nearness between sets, a generalization of roughness, is matter of closeness in the values of the perceptual features in the description vector. The nearness relation is defined in terms of similarity of descriptions and the added tolerance relation defined in terms of difference in values of selected features.

The fact that these set-theoretic treatments are motivated by cognitive interests in perceptual systems and categorization places their interpretation squarely in the epistemic camp. However, it is also compatible with the different kinds of objectivity of the formal structure and the material basis of (relational) perceptual properties and control. The material kind of objectivity is instantiated by the physical models of perceptual systems considered as psycho-physical mappings, by for instance, neural networks and technological systems. Even relations of indiscernibility can receive an objective interpretation in terms of its realization in identical physical states—whether brain states, behavioral consequences or technological expression. As part of the complex cognitive environment, however, the application of these formal methods to image analysis and processing requires the application of empirical standards and judgments—which in turn may receive ulterior objective material modeling.

In general, epistemic interpretations of approximation extend to several related methodological contexts: for instance, the empirical application of quantities in experimental situations (context 4) in which approximation becomes a matter of measurement (context 3) used as evidence (context 5) as part of a process of inquiry aiming at increasingly closer approximation (context 9) in which the measurement and evidence assessments include pragmatic considerations and boundaries (context 7) typically of the form 'close enough for the given purpose' or 'in the given context.'[12] Distances that characterize near-collision events between the Earth and an asteroid are vastly greater than ones in near-collisions between cars. Evidence must be understood generally as relevant to establishing both a hypothesis and a fact, that is the criterion for the adequate application of a concept or label—a constitutive or semantic hypothesis.

In those situations fuzziness and accuracy combine as two dimensions of approximation, with their respective metrics and assessment standards. Relative to a fixed target and a margin of error, we can assess the accuracy of a measurement and the evidentiary support it provides for some prediction. But the accuracy depends on a prior standard of precision. I distinguish between two kinds of precision. It may be conceptual, or qualitative, even if it applies to a precise quantity, as in whether a temperature of 65° makes something warm or a height of 6 feet makes a person tall. In such a case we require two quantitative strategies that enable an assessment of accuracy. First we can establish a degree of membership in the relevant set expressing the category, or quality, in question. Then we impose a filter

---

[12]Ibid.

or boundary based on contextual and pragmatic considerations; this second step also helps establish a precise quantification of the quality.

The second kind of precision concerns the quantification in the second step in a measurement determinations, since a variety of values qualify within a range as an equivalent class relative to the empirical fact of the matter.[13] We associate this uncertainty with further aspects, the sensitivity of a measurement instrument and the units of measurement that help apply the numerical scale. Only after we have fixed the precision of the potential representation we can apply it in such a way that we can determine its degree of accuracy by statistical and theoretical standards.

In conclusion, applying the view of approximation I have delineated helps bring out in more detail the relational dimension of fuzziness. More specifically, it represents the centered character of practices of formal treatment and technological control, guided by standards of maximum and minimum fuzziness and targets of minimal fuzziness. I describe other dimensions of this and other practices in the next chapter.

References

1. Siegel, S. (2010). *The contents of visual experience*. New York: Oxford University Press.
2. Feyerabend, P. K. (1975). *Against method*. London: Verso.
3. Kline, S. J. (1981). *Similitude and approximation theory*. Stanford, CA: Stanford University Press.
4. Bachelard, G. (1927). *Essai sur la Connaissance Approchée*. Paris: Vrin.
5. Cat, J. (2015). An informal meditation on empiricism and approximation in fuzzy logic and fuzzy set theory: Between subjectivity and normativity. In R. Seising, E. Trillas, & J. Kacprzyk (Eds.), *Fuzzy logic: Towards the future* (pp. 179–234). Berlin: Springer.
6. Krantz, D., Luce, D., Suppes, P., & Tversky, A. (1971). *Foundations of measurement* (Vol. 3). New York: Dover.
7. Cat, J. (2016). The performative construction of natural kinds: Mathematical application as practice. In C. Kendig (Ed.), *Natural kinds and classification in scientific practice* (pp. 87–105). Abingdon: Routledge.
8. Falguera, J. L., & Peleteiro, S. (2014). Percepción y justificación, legitimación o sustento? *ESTYLF 2014, Libro de Actas*. Zaragoza: Universidad de Zaragoza, pp. 441–446.
9. Hopkins, Ch. (1998). *Picture, image, and experience*. Cambridge: Cambridge University Press.
10. Pawlak, Z. (1982). Rough sets. *International Journal for Computing and Information Science, 11*, 341–356.
11. Duhem, P. (1906). *La Théorie Phyique: Son Objet et Sa Structure*. Paris: Vrin.

---

[13]In this sense a measurement may be quantitatively imprecise or fuzzy before it can be determined to be inaccurate by some margin; see Duhem [11]. Duhem appeals to the linguistic semantic of symbolic denotation to claim that predicate standing for qualities denote like a symbol, but do not picture empirical facts and that quantitative theoretical laws can be neither true nor false, only fuzzy approximations.

Chapter 21
Pictorial Vagueness as Scientific Practice: Picture-Making and the Mathematical Practice of Fuzzy Categorization

Vagueness is practice and is in mathematical practice too. In this chapter I take stock of assumptions and conclusions I have presented in previous chapters and I examine them further. To place fuzziness at the center of a philosophical account of vagueness is to place more than subjective uncertainty or an objective property in an alternative to epistemic accounts; it is to place also an objective activity and a practice of mathematics and of human and machine categorization.[1] Fuzziness represents them and its empirical and technical applications aim to control and exploit it. If vagueness is best understood objectively, from the standpoint of fuzziness, objectivity must be understood more broadly than an ontic commitment to the reality of autonomous preexisting properties in the world. This extends to the case defended here of vagueness or uncertainty in pictures.

Fuzziness in pictures gets its diverse and specific significance in the contexts of cognitive activities and practices. The extension connects the philosophical debates to the scientific practice by extending the philosophical discussion of vagueness; in turn it provides the scientific practice with additional understanding and guidance. Researches in the technical and the philosophical frameworks surrounding the formulation of fuzziness contribute to each other.

In art, perception and the application of mathematics, fuzziness plays an important role in activities engaging pictures, from making to using. I have mentioned different practices endorsing different values and purposes for visual fuzziness—aesthetic, cognitive or practical. Mathematics is among them; it is indeed as much a medium as are paint and language, ready to be applied and practiced with. In mathematics and its technological applications vagueness is a practice. It is in the linguistic symbolic case and I suggest that it is too in the pictorial case.

---

[1]Smith [1].

© Springer International Publishing AG 2017

J. Cat, *Fuzzy Pictures as Philosophical Problem and Scientific Practice*,
Studies in Fuzziness and Soft Computing 348,
DOI 10.1007/978-3-319-47190-7_21

As I have mentioned, in its inception, Zadeh's motivation to develop a generalization of set theory included a number of formal attempts to treat ordinary and scientific systems and practices, from the mathematical representation of biological systems to the quantification of ordinary word usage and of truth in multivalued logic.[2] Symbolic rules associated with fuzzy categorization and reasoning provided the tools for a successful application of the formalism in technological systems.

Mathematics may be classed, then, with art and ordinary linguistic expression in the ways presented above. In particular, in mathematics and its technological applications vagueness is a cognitive activity and a community practice—in the scientific case, a disciplinary practice. Here I have adopted the view that a practice is purposeful and regulated, relying on the exercise of a skill; its normative dimension is expressed through the reliability of habits within a community.[3] On these assumptions, activity and practice are inseparable. Fuzzy set theory developed into a disciplinary practice for the empirical modeling and controlling vagueness first in two ordinary cognitive activities, informal categorization and linguistic representation and informal, so-called approximate reasoning.

Categorization and its role in modeling vagueness are a cognitive and community practice that we can represent with technical notions of fuzziness. It is a matter of contextual and practical objectivity, involving formal construction, subjective judgments, decisions, rules, purposes and particular standards. For instance, a key part of this partially objective methodology often involves assigning measures of similarity.[4] The same activities constitute the practice of representation, evaluation and control of fuzziness in images.

The application of fuzziness to the analysis of images rests on standards of construction and evaluation of models and procedures—through algorithms—for categorization—classification—of images, their processing and their further identification. Construction standards include formal constraints from the conceptual framework and empirical assumptions about categorization. Evaluation standards include formal and visual pictorial judgment.

The aim of categorization through clustering algorithms already involves the application of mathematical conditions expressing values that are both contextual and conceptual. For instance, in unimodal, grey MRI imaging, the choice of initial centroids is the pragmatic and contextual choice of standard values that makes possible to run the fuzzy c-mean clustering algorithm over sets of grey level values on a unimodal histogram. The choice incorporates the cognitive aim to maximize measures of class separability; one measure for the separation is a threshold value function of the average difference between the highest and second-highest membership value of each cluster object over the total number of objects, e.g., pixels, as the data set. Pixels can be then associated with vectors composed of a series of components representing individual and statistical feature values.

---

[2]Zadeh [2, 3], McNeill and Freiberger [4], Seising [5], Cat [6, 7].
[3]Turner [8].
[4]Cat [6].

Segmentation of clusters rely on what I have called IC-EC rules based on conceptual, empirical and pragmatic assumptions about the distinction between object and background and about the border on which the distinction rests. The border is given a formal relational characterization at each point, e.g., set of maximum differences of average grey levels in pairs of horizontally and vertically adjacent neighborhoods. The characterization is based on a measure of edge value in the form of grey level and level rate of change.[5] In the rough-fuzzy approach, clusters are represented by a centroid, a crisp lower approximation and a fuzzy border.

More statistical intrinsic properties include so-called textural features such as measures of the probability of co-occurrence of a pair of grey levels separated by a given distance at a given angle. Such measures take the form of matrices or probability density functions for gray-level co-occurrence. Then, attention to uniformity of texture values over a region facilitates the task of segmentation into homogeneous blocks at the expense of resolution in terms of smallest size of homogeneous sub-images. The value of textural features consists in mediating between IC and EC, but the identification of objects or regions of interests requires low-level IC-EC constraints that establish the priority and interpretation of resolution, say, over border contrast. In algorithms for segmenting, for instance, MR brain images, a standard of acceptable resolution will be necessary.

Next, a process of evaluation of the constructed images follows the application of techniques of analysis by segmentation. Formal indices represent different orderings of degrees of clustering performance according to specific aims and standards. These are typically motivated and constrained by contextual empirical intuitions and by a variety of formal measures: the Davies-Bouldin index, the Dunn index, the beta index, mean squared error, in-class and between-class variance, etc. For instance, for a given distribution of values associated with a feature (or a feature vector representing a set of them) and the selection of initial centroids for clustering the beta index maximizes homogeneity in a segmented region as the ratio of within-cluster variation to total variation; and the Dunn index and the Davies-Bouldin index endorse different relations between a minimum within-cluster distance and a maximum between-cluster separation. In addition, there are performance constraints set by practical desiderata such as computing time.[6] Then the optimization algorithm for segmenting and classifying images is tested against different performance indices.

Also, the motivation for supplementing fuzzy-set algorithms with rough-set ones is based on two methodological considerations of empirical adequacy. One is that fuzzy membership values help model overlapping partitions of clusters; the weakness lies in the outcome of c-mean clustering algorithms, in which the computed membership values of data fail to match the assigned values and become indiscernible from environmental noise, without any criterion for distinguishing

---

[5]Maji and Pal [9], 2.11.
[6]Ibid., 2.8.

between the two. Both shortcomings are expressions of constraints on method-ological practices.

An additional consideration is explanatory; the empirical inaccuracy of the computed membership values fails to support the algorithm as their explanation.[7] The application of rough sets, in terms of rough resemblance, supplement the role of fuzzy sets by contributing to class definition, another epistemic value or cognitive goal.

For the application of formal expressions and rules external constraints in the form of standards and decisions about them are needed. Formal methods aim to make automatic the process of formation of visual and qualitative determinations—ex., interpretation, recognition, discrimination or classification—by human opera-tors. To that effect, the automatic character of the methods relies on formal preci-sion in modeling uncertainty and classification. For instance, based on an optimized image, additional PIC-PEC links will introduce additional schema of categorization and analysis—anatomical, pathological, etc.—relevant to some purpose in the context at hand.

But formal precision does not guarantee correctness, error-free qualitative accuracy of computation, or even eliminate conceptual vagueness. So the judgment of human operators is often considered. The clustering standards and so-called certainty grades in the weighted patterns that help optimize the classification rule base require external human (expert) judgment and choice. The application of pictorial standards already includes a role for the human factor. Pictorial analysis includes two elements: so-called ground-truth images (the base standard) and supervising human recognition judgment. The information in ground-truth images may be utilized through formal relations. Supplementary human judgment includes so-called human hand-labeling of standards, definition of classes and validation of classes and classification of specific images.

By now it should have become clear that the meaning of vagueness in pictures in terms of fuzziness of their properties/categorizations is tied to sets of practices in which it is embedded. They are characterized by their aims, values and standards that help apply them. The objectivity of fuzziness as represented by the set-theoretic concept is not unavoidably reduced to the objectivity of instantiated properties. Minimally, attributed properties in that sense only reflect the context of application of mathematical formalism; therefore, their claimed empirical objectivity must include the empirical objectivity of cognitive practices and their methods. This extended objectivity incorporates the scientific interpretation of fuzziness while blurring the sharp ontological distinction between opposing objective and epistemic accounts of vagueness in philosophical debates.

The scientific concept of fuzziness is inseparable from practices of formal and empirical application; they take the form of diverse and contextual practices of construction and evaluation and control. As a result, fuzzy representation is inseparable from practices of construction, evaluation and intervention. While this

---

[7]Maji and Pal [9], 2.1–2.

feature isn't distinctive of the formal representation of vagueness in pictures, it provides the basis of the distinctive ways in which cognitive practices of representation and reasoning are possible with pictures.

References

1. Smith, N. J. J. (2010). *Vagueness and degrees of truth*. Oxford: Oxford University Press.
2. Zadeh, L. A. (1962). From circuit theory to system theory. *Proceedings of Institution of Radio Engineers, 50*, 856–865.
3. Zadeh, L. A. (1965). Fuzzy sets. *Information and Control, 201*, 240–256.
4. McNeil, D., & Freiberger, P. (1993). *Fuzzy logic*. New York: Touchstone.
5. Seising, R. (2007). *Fuzzification of systems: The genesis of fuzzy set theory and its initial applications*. New York: Springer.
6. Cat, J. (2015). An informal meditation on empiricism and approximation in fuzzy logic and fuzzy set theory: Between subjectivity and normativity. In R. Seising, E. Trillas, & J. Kacprzyk (Eds.), *Fuzzy logic: Towards the future* (pp. 179–234). Berlin: Springer.
7. Cat, J. (2016). The performative construction of natural kinds: mathematical application as practice. In C. Kendig (Ed.), *Natural kinds and classification in scientific practice* (pp. 87–105). Abingdon and New York: Routledge.
8. Turner, S. P. (2010). *Explaining the normative*. Malden, MA: Polity Press.
9. Maji, P. & Pal, S. K. (2010). Rough-fuzzy clustering algorithm for segmentation of brain MR images. In: Pal, S. K., Peters, J. F. pp 2.1–21.

Chapter 22
Conclusion

In this book I have taken up a problem in philosophy, making sense of vagueness, extended to pictures and tracked one particular solution into the world of applied mathematical science, fuzzy set theory. My interest in this problem is twofold.[1] There is the philosophical problem itself, and then there is the problem's role of laying down lines of communication across divides between philosophical and scientific practices. From some disciplinary distance, these are practices that can share similarly formal and conceptual preoccupations, also an empirical interest in actual human practices such as representation and cognition. The philosophical problem emerges out of an interest in semantic practices in the use of symbolic language, especially in representation and reasoning. Science offers its own symbolic, mathematical, practices to represent vagueness and to apply it in empirical representation of empirical phenomena. Here is a case in which science, mathematics, solves a problem for philosophy and philosophy retrieves it back for further analysis and use. I have also identified explicit problems and sources adopted from more philosophical projects. Science, then, doesn't conflict with philosophy, nor does it replace it.

What is this further use of science? I have argued that the further philosophical exploration of the scientific account yields a triple result in the form of inseparable solutions to three general problems: vagueness—fuzziness, uncertainty or imprecision—, pictorial representation and the application of mathematics in scientific practice.

Specifically, understanding the nature of the solution in terms of categorization provides the opportunity to extend the problem to the domain of pictorial practices and, again, to consider a similar solution from science. I have argued also that to understand the scope of that solution involves understanding scientific as well as artistic practices and cognitive activities. Vagueness in practice finds its rich contextual significance attached to diverse sets of activities, values and purposes. I have emphasized the diversity of uses and meanings, especially in technical treatments

---

[1]I have expressed similar views in relation to causation in Cat [1].

© Springer International Publishing AG 2017
J. Cat, *Fuzzy Pictures as Philosophical Problem and Scientific Practice*,
Studies in Fuzziness and Soft Computing 348,
DOI 10.1007/978-3-319-47190-7_22

within set theory. Scientific practices of empirical representation and technological control appear inseparable. In addition, the application of fuzziness to understanding and controlling categorization and reasoning is itself an instance of a practice of categorization and reasoning in the application of mathematics. To accommodate the full extent of the scientific solution to the problems of linguistic and pictorial in terms of categorization, I have argued that it helps adopting more encompassing notions of categorization and empirical objectivity; but they do not eliminate the role of the subjective, centered dimension.

Within this extended framework, the case of pictures shows distinctive complications and challenges, also opportunities; for instance, in the philosophical and scientific understanding and use of vagueness and pictorial representation. The practices of pictorial representation and thinking present themselves as growth opportunities for scientific representation and control. But modeling such possibilities requires rejecting as self-evident the goal of minimizing fuzziness; this is a matter set by the meanings, standards and possibilities available within each context of practice. It requires also accommodating the complex conditions of integration of pictorial and symbolic uses of visual designs, and operating outside the boundaries set by the restrictive semantic standard of symbolic truth.

Reference

1. Cat, J. (2006). On fuzzy empiricism and fuzzy-set models of causality: What is all the fuzz about? *Philosophy of Science, 73*(1), 26–41.

Printed in the United States
By Bookmasters